LIBRARY IN A BOOK

BIOTECHNOLOGY AND GENETIC ENGINEERING

Revised Edition

Lisa Yount

Facts On File, Inc.

To the scientists who will create the "Biotech Century"
and the children who will have to deal with it

BIOTECHNOLOGY AND GENETIC ENGINEERING,
REVISED EDITION

Facts On File, Inc.
132 West 31st Street
New York NY 10001

Library of Congress Cataloging-in-Publication Data
Yount, Lisa.
Biotechnology and genetic engineering / Lisa Yount.—Rev. ed.
p. cm.
Includes bibliographical references and index.
ISBN 0-8160-5059-7
1. Biotechnology—Social aspects. 2. Genetic engineering—Social aspects. I. Title.
TP248.23.Y684 2004
303.48′3—dc22 2003064223

Facts On File books are available at special discounts when purchased in bulk quantities for businesses, associations, institutions, or sales promotions. Please call our Special Sales Department in New York at (212) 967-8800 or (800) 322-8755.

You can find Facts On File on the World Wide Web at http://www.factsonfile.com.

Text design by Ron Monteleone

Printed in the United States of America

MP Hermitage 10 9 8 7 6 5 4 3 2 1

This book is printed on acid-free paper.

CONTENTS

PART I
OVERVIEW OF THE TOPIC

PART II
GUIDE TO FURTHER RESEARCH

PART III
APPENDICES

PART I

OVERVIEW OF THE TOPIC

CHAPTER 1

ISSUES IN BIOTECHNOLOGY AND GENETIC ENGINEERING

In April 2003, the International Human Genome Sequencing Consortium announced completion of the Human Genome Project, a huge effort—lasting more than a decade—to obtain the complete readout of the code, some 3 billion "letters" worth, that contains all the inherited information specifying the characteristics of a human being. *Genomics & Genetics Weekly* called this work "one of the most ambitious scientific undertakings of all time."[1] Appropriately, April 2003 also marked the 50th anniversary of the publication of the structure of DNA, the key discovery that allowed scientists to learn the nature of the genetic code and the way it is transmitted, and approximately the 30th anniversary of the birth of genetic engineering, which enables researchers to change the genetic makeup of living things.

As great an achievement as completion of the Human Genome Project is, the researchers involved in it know that they are only at the beginning of their work. Possessing a blueprint is not the same as having a finished building. Scientists may have the "text" for all of humanity's genes, but they do not know what most of those genes do nor how they interact with each other and the environment to produce a living, changing individual. Genetic science and the biotechnology industry it has spawned are still in their infancy, surrounded by a cloud of hopes and fears. Like many of the hopes and fears surrounding space exploration, some of these will never come to pass. Nonetheless, few people doubt that sooner or later genetic engineering and biotechnology will produce massive changes in the way people live, their health, their environment, and, perhaps, in the very nature of humanity. Supporters and critics alike expect the 21st century to be what one critic, Jeremy Rifkin, has dubbed "the biotech century."[2]

The Human Genome Project, along with other achievements in genetics, genetic engineering, and biotechnology, raises many social, legal, and ethical issues. When scientists and businesspeople tinker with the very stuff of life, everyone is potentially affected. This book focuses on five areas in which such issues are especially sensitive: health and environmental safety, particularly in regard to agricultural biotechnology; the patenting of living things, tissues, cells, and genes; DNA testing for identification in criminal cases; genetic testing for susceptibility to disease and the possibility of genetic discrimination in employment and insurance; and alteration of human genes, including new gene-based treatments for disease and the possible cloning of human beings. The remainder of this introduction will describe the issues in these areas, the technologies that prompted them, and the way governments, courts, and the public have perceived and reacted to them.

CRACKING THE CODE OF LIFE

In some senses, biotechnology—the use or alteration of other living things in processes that benefit humankind—is almost as old as humanity itself. Biotechnology certainly has existed as long as people have raised crops and domesticated animals, practices they began some 10,000 years ago. Although they were unaware that they were doing so, people used microorganisms in the processes of making bread, cheese, alcoholic drinks, and tanned leather. When people began sowing only the seeds from the best (most productive, most disease-resistant, and so on) of the previous year's crop plants, and breeding only the strongest or most productive cattle and other domestic animals, they also, again unknowingly, altered the genes of living things to better meet their needs. When they learned that they could breed certain plants from cuttings, they discovered cloning (the word *clone* means "twig" in Greek).

Only in the 19th century did people begin to learn what was really happening during these technological efforts. Starting in the late 1850s, French chemist Louis Pasteur showed that living yeast or other microorganisms were necessary for the fermentation that produced wine, beer, and other products. More important to the genetic side of modern biotechnology, in 1866 an obscure Austrian monk named Gregor Mendel published a paper describing the mathematical rules of inheritance of physical traits. He had worked out these rules by breeding and observing pea plants in his monastery garden. Mendel's rules showed the statistical pattern in which such characteristics as height and seed color were inherited by the hybrid offspring of two plants that differed from each other in these characteristics.

The form of a characteristic that an offspring would receive was determined by what Mendel called factors, one of which (for each characteristic) the offspring inherited from its male parent and one from its female parent. Some factors were more powerful, or dominant, than others. If a plant inherited one dominant factor and one weaker (recessive) factor for a trait, the plant would exhibit the dominant form of the trait.

Few knew of Mendel's work at the time, let alone realized that it went far toward explaining the theory of Charles Darwin, described in his *On the Origin of Species* (first published in 1859), which stated that nature acted like traditional plant and animal breeders to select and preserve the inherited characteristics that helped species survive in their environment. Darwin called his theory "evolution by natural selection."

Earlier in the 19th century, Matthias Schleiden and Theodor Schwann had proposed that microscopic bodies called cells were the basic units from which all living things were formed. Improvements in microscopes allowed scientists to begin exploring the cell's inner structures toward the end of the century. Mendel's rules of heredity and the study of cells converged in the early 1900s, when Mendel's forgotten work was rediscovered and scientists began speculating that his "factors" were somehow contained in threadlike chromosomes, or "colored bodies," which German biologist Walther Flemming had discovered in the nuclei (central bodies) of cells in 1875. (All cells except those of bacteria and other so-called prokaryotes contain nuclei and chromosomes.)

Drawing on experiments with fruit flies, Thomas Hunt Morgan of Columbia University and his coworkers proved the link between chromosomes and heredity in 1910. By then Mendel's factors had taken on the new name of *genes*, and the science of genetics had been born. In the decades that followed, Morgan's group and others began to make maps of chromosomes showing approximately where the genes that determined certain characteristics were located.

Geneticists assumed that inherited information must be coded somehow in the structure of either proteins or nucleic acids, the only types of chemical in the chromosomes. At first they thought proteins the more likely information carriers because these chemicals were known to have a complex structure. In 1944, however, Oswald Avery and his coworkers showed that a harmless strain of certain bacteria became capable of causing disease when they took up deoxyribonucleic acid (DNA) from a disease-causing strain of the same bacteria. The bacteria's descendants retained the new disease-causing ability, indicating that information for producing this trait must have been contained in the DNA. Avery's work turned the genetic spotlight from proteins to nucleic acids, especially DNA.

At that time DNA's chemical composition was known, but key aspects of its structure were not. Biologists knew that the DNA molecule had a long "backbone" composed of smaller molecules of phosphate and a sugar called deoxyribose. Attached to this backbone were many units of four kinds of small molecules called bases: adenine (A), thymine (T), guanine (G), and cytosine (C). X-ray crystallography photographs suggested that the DNA molecule had the general shape of a helix, or coil. However, no one knew exactly how many backbone chains the molecule had, how the bases were attached to them, or how all these units were packed within the molecule.

Geneticists and biochemists in the late 1940s and early 1950s increasingly realized that the structure of DNA must hold the secret of its ability to transmit biological instructions from one generation to the next, and several groups of scientists began trying to figure out that structure. The ones who first succeeded were a young American, James Watson, and a somewhat older Englishman, Francis Crick, who worked together at England's prestigious Cambridge University. Drawing on molecular models and X-ray photographs, they concluded that each DNA molecule contained two sugar-phosphate backbones that entwined one another in a corkscrew shape that came to be known as the "double helix." Pairs of bases stretched between the two backbones like rungs on a twisted ladder. A molecule of adenine always paired with one of thymine, and cytosine paired with guanine.

Watson and Crick published their groundbreaking account of DNA's structure in the British science journal *Nature* on April 25, 1953. Five weeks later, they published a second paper describing how this structure could allow DNA to reproduce itself so that a complete copy of hereditary instructions was given to each of the "daughter" cells when a cell divided. The hydrogen bonds that joined the bases in each pair were weak, Watson and Crick explained, so the bonds might dissolve as the cell prepared to divide. Each DNA molecule would thus split down its length, like a zipper unzipping. Each base could then attract a molecule of the complementary base, bearing with it a sugar-phosphate "backbone" unit, from free-floating materials in the cell, and the bonds joining the bases would re-form. The result would be two separate DNA molecules identical to the original one. This theory was proved correct in 1958.

DNA's structure also revealed the "code" in which inherited information is stored. Molecular biologists had known since 1941 that genes contain instructions for making proteins, the complex chemicals that do most of the work of the cell. Indeed, scientists had come to define a gene as the portion of a DNA molecule that carries the instructions for making one protein. (This proved to be an oversimplification. Some genes make molecules of ribonucleic acid [RNA] or modify the activity of other genes.) The most likely candidate for the code in which these instructions were carried was the order of the bases in the DNA molecule. DNA has only four kinds of bases, however, whereas proteins consist of 20 types of simpler substances called amino acids. In 1961, Crick suggested that the "code letter" specifying a particular type of amino acid could be a sequence of three bases. The 64 (4 × 4 × 4) possible triads were more than enough to specify all of the amino acids.

During the next five years, several groups of scientists painstakingly deciphered this genetic code, determining which amino acid each of the 64 triads stood for. (Some amino acids could be represented by any of several different triads.) Meanwhile, others worked out the procedure by which information in the DNA code was translated into protein. They discovered that the DNA is copied into a form of RNA, a related chemical that differs from DNA only in having uracil rather than thymine as one of its four bases. Unlike DNA, RNA can leave the cell's nucleus and move out into the cytoplasm, the main substance of the cell body. There the RNA guides the assembly of amino acids, in the order specified by the original DNA, into a protein molecule.

THE BIRTH OF GENETIC ENGINEERING

Building on these basic discoveries, scientists in the early 1970s developed the ability to change genes instead of merely deciphering them. Genetic engineering—the direct alteration of genes or transfer of genes from one type of organism to another—works because (except in the case of a few viruses with genes made of RNA rather than DNA) all genes have the same basic structure and work in the same way. Thus, they are in a sense interchangeable. Once a gene is placed in a cell's genome, or collection of genes, that cell becomes able to make the protein for which the gene codes, regardless of whether it ever made that substance in nature. As an example, bacteria or cattle can be made to produce human hormones by inserting the appropriate human genes into them.

Genetic engineering began in the laboratory of Paul Berg, a Stanford University biochemist, in the winter of 1972–73. Berg removed a gene from SV40 (Simian Virus 40), a monkey virus that could cause cancer in mice. Through laborious chemical manipulation, he attached to it a short piece of single-stranded DNA. He then opened up the small, circular genome of another virus, lambda, and attached a chain of single-stranded DNA with a base sequence complementary to that on the added SV40 piece to one of the lambda genome's open ends. He termed these single-stranded chains "sticky ends" because any such chain will attach itself to another strand with a complementary base sequence (a sequence in which all the bases in the second chain will pair with those in the first). For example, a chain with the sequence T-A-G-C will attach itself to another with the sequence A-T-C-G. Taking advantage of this "stickiness," Berg spliced the gene from SV40 into the lambda genome. This was the first production of recombinant DNA, or DNA into which other DNA from a different type of organism has been inserted.

Berg's technique for gene splicing would have been hard to apply on a mass scale, but the same was not true of a second technique developed shortly afterward by another Stanford scientist, Stanley Cohen, and Herbert Boyer, from the University of California at San Francisco. Appropriately, the idea for this technique—which would allow researchers to (in effect) slice and dice genes, sandwich them in any order, and pack them "to go"—was born in a delicatessen. Cohen and Boyer met at this delicatessen one evening in November 1972 after attending a scientific meeting in Honolulu, Hawaii. As they discussed their work, each discovered that the other's research held a missing piece of his own puzzle.

Unlike Berg, Cohen and Boyer were working with bacteria. A bacterium's genome is large compared to that of a virus, containing an average of 2,500 genes rather than a few hundred, although it is far smaller than those of cells with nuclei (the human genome, for instance, probably contains between 25,000 and 60,000 genes). Most of a bacterium's genes are carried in a single large ring of DNA, but some are contained in smaller rings called plasmids, which the bacteria can exchange in several ways. Cohen had invented a technique for removing plasmids from one bacterial cell and inserting them into another.

Boyer, for his part, was working with bacteria called *Escherichia coli* (*E. coli* for short), which commonly and usually harmlessly live in the human intestine. He was studying enzymes (proteins that catalyze chemical reactions in living cells) called restriction endonucleases, which *E. coli* and some other bacteria produce as a defense against viruses. These enzymes snip DNA into pieces. Each restriction enzyme breaks a DNA molecule apart at every spot where it encounters a particular sequence of bases, and the process conveniently produces the same kind of single-stranded "sticky ends" that Berg had so painstakingly created with his chemicals. Any two pieces of DNA cut by the same restriction enzyme, therefore, can be joined together, even if they come from very different species. Different restriction enzymes act on different base sequences. Hundreds of restriction enzymes are known today, and they have become basic tools of gene splicing. So have the ligases, a type of enzyme that both Boyer and Berg used to weld together their spliced DNA fragments.

Chatting over sandwiches, Cohen and Boyer realized that if they applied Boyer's *E. coli* restriction enzyme, EcoR1, to Cohen's plasmids, they would have a way to cut open a plasmid from one species of bacteria and, with the help of a ligase, splice it onto a plasmid from another species. They proceeded to try this in spring 1973. One of their two plasmids contained a gene that conferred resistance to a certain antibiotic. Boyer and Cohen allowed a type of bacterium that was not naturally resistant to that antibiotic to take up the combined plasmids, then placed the bacteria in a culture medium containing the antibiotic. Some of the bacteria survived, which showed that the resistance gene in the engineered plasmids was still making its signature protein. Later, the two scientists spliced some DNA from a frog into an *E. coli* plasmid and proved that it functioned there as well.

In addition to plasmids, Boyer, Cohen, and other scientists were soon using viruses (such as lambda, which infects bacteria) as vectors, or transmission agents, for inserting foreign genes into bacteria and, later, plant and animal cells. Viruses—nature's own genetic engineers—reproduce by inserting their genes into cells, which then reproduce the viral genomes along with their own; when scientists add genes to the viruses, these are inserted and reproduced as well. As one scientist commented when he heard about the new techniques, "Now we can put together any DNA we want to."[3] The gene age had begun.

Although traditional biotechnology continues, the term has now become almost synonymous with genetic engineering. Together, biotechnology and genetic engineering represent not only a scientific field but an industry. In the three decades since the "gene deli" opened, companies based on the technology of decoding and altering genes have multiplied, seemingly as fast as the bacteria in their founders' test tubes. A report from British business advice firm Ernst & Young provided the following figures for 2001.

To be sure, biotechnology so far has not been the profit maker that its investors have hoped for. San Francisco financier Steve Burrill claimed in early 2003 that publicly traded U.S. biotechnology firms had in current market capitalization about the same total amount of money that investors had put in—about $220 billion. In other words, said Burrill, "We haven't as an industry returned any money to investors."[4] Because of this performance as well as an overall poor economy, many investors have become leery of the

Country/Area	Revenues	Net Income	Employees	Companies
United States	$25.3 billion	–$4,799 million	141,000	1,115 private, 342 public
Europe	$7.5 billion	–$608 million	34,180	1,775 private, 104 public
Canada	$1.0 billion	–$507 million	7,005	331 private, 85 public
Asia/Pacific Rim	$1.0 billion	–$19 million	6,518	441 private, 91 public

field. "At the moment, there is virtually no money for new [biotech] ventures," an *Economist* article reported in March 2003. Nonetheless, the article went on to say that giving up on biotechnology as an investment "is to confuse short-term problems with long-term potential."[5] Burrill said that despite current problems, "the longer term outlook for biotech is excellent."[6]

Agricultural Biotechnology: Safety Issues

Concerns about the safety of genetic engineering experiments and genetically altered organisms began even before genetic engineering itself. In 1971, when Robert Pollack, a geneticist at Cold Spring Harbor Laboratory on Long Island, New York, learned that Paul Berg planned to allow lambda containing SV40 genes to infect *E. coli*, he telephoned Berg to express his concern over the possible consequences of inserting genes from a cancer-causing virus into one that could infect bacteria that live in the human intestine. It might be possible, Pollack said, for bacteria containing cancer genes to escape from the laboratory and infect people. Berg decided that Pollack might be right. He transferred genes from SV40 to lambda, but he did not allow the altered lambda to infect bacteria.

Other scientists, including Boyer and Cohen, came to share Berg's and Pollack's uneasiness. As some of them pointed out in a discussion at the Gordon Conference on Nucleic Acids in mid-1973, *E. coli*, the "workhorse" bacterium that many experimenters in the new field used, not only could infect humans itself but could exchange plasmids with other bacteria, including ones that cause human disease. The result of this and similar discussions was two letters to *Science*, the prestigious journal of the American Association for the Advancement of Science, warning of the possible dangers of recombinant DNA research. The second letter, signed by top scientists in the field, including Berg, Boyer, and Cohen, and published in the July 26, 1974 issue of *Science*, was the more detailed of the two. It recommended halting several types of recombinant experiments, including those involving genes for antibiotic resistance or tumor formation, until possible hazards could be more thoroughly evaluated and guidelines for safe procedure could be established. This was the first time a group of scientists had voluntarily proposed halting a certain type of experiment because of possible dangers.

Fear of possible harm from accidental release of genetically engineered microorganisms brought 140 geneticists and molecular biologists together for a historic conference at Asilomar, a seaside retreat center in central California, in February 1975. Michael Rogers, a journalist writing in *Rolling Stone*, called it "the Pandora's Box Congress."[7] The Asilomar scientists concluded that most work on construction of recombinant DNA molecules should proceed, provided that appropriate safeguards were employed, but they agreed not to perform some potentially risky types of experiments until new laboratories with special safeguards for containing potentially dangerous microbes were built. Then and later, the Asilomar scientists were praised for their restraint. Cynical observers and some of the scientists themselves, however, pointed out that their actions were at least partly an attempt to avoid public criticism and government regulation so that they could maintain control of this exciting new area of research.

The Asilomar guidelines became the blueprint for the first federal government regulations on genetic engineering. These regulations, issued in June 1976, were drafted by the Recombinant DNA Advisory Committee (RAC), a panel of molecular biologists and other scientists that the National Institutes of Health (NIH) had established in October 1974. The RAC had to approve all recombinant DNA experiments funded by the NIH, which at the time meant most of such experiments in the United States. The RAC/NIH regulations were binding on all federally funded research, and most private (usually university-based) researchers agreed to abide by them as well. About this same time, Britain published similar guidelines and established a government oversight group much like the RAC. The European Molecular Biology Organization (EMBO) recommended that researchers in other European countries follow either the U.S. or the British guidelines.

Worry about the possible escape of genetically engineered "superbugs" was not limited to scientists. Congress held its first hearings on the issue in April 1975. Sensationalist media articles, often blending fact with speculation, stirred up the public's fears. Critics, including several environmental groups, complained that the NIH guidelines were not strict enough, accused RAC scientists of conflict of interest (many of them carried out the kind of research they were supposed to regulate), and demanded that ethicists and environmentalists have a hand in drawing up future regulations. Debate about the new science was so intense in the mid-1970s that Rockefeller University professor Norton Zinder called it the "Recombinant DNA War."[8]

Friends and foes alike were concerned that the NIH had little, if any, legal authority to regulate research other than that which it funded. Indeed, a federal interagency committee reported to U.S. Health, Education, and Welfare Secretary Joseph Califano on March 15, 1977, that, as far as it could tell, no existing federal agency had the authority to handle all problems raised by recombinant DNA research. Sixteen bills to regulate recombinant DNA research were introduced into Congress that year in attempts to remedy this situation, but none became law, partly because public concern had waned by then and partly because most scientists in the field had come to fear restrictive legislation even more than runaway bacteria. They lobbied against the regulatory bills and convinced the legislators that most recombinant DNA experiments presented no threat to public health.

As time passed and no evidence of actual harm emerged, the NIH relaxed its guidelines, eventually exempting most experiments from regulation. By the early 1980s, most scientific and regulatory bodies had ceased to regard the process of genetic engineering as inherently risky. Controversy arose again in mid-decade, however, when Steven Lindow, a plant pathologist at the University of California, Berkeley, proposed to release genetically engineered organisms—a common and harmless type of bacteria that he had engineered to make citrus and certain other plants resistant to frost damage—into the environment for the first time, as part of a small field test. Jeremy Rifkin, director of the Foundation on Economic Trends and a foe of genetic engineering since the "recombinant DNA war" days, opposed the test, calling it "ecological roulette," and several environmental groups joined him.[9] They succeeded in delaying the test for four years, but when Lindow finally carried it out on April 24, 1987, no detectable environmental effects occurred.

By this time, genetic engineering was being applied to plants and animals as well as bacteria, and the RAC was no longer the only federal body regulating it. In June 1986, the U.S. Office of Science and Technology Policy had published the Coordinated Framework for Regulation of Biotechnology, which divided regulation of genetic engineering research and products among five agencies: the NIH, the National Science Foundation (NSF), the U.S. Department of Agriculture (USDA), the Environmental Protection Agency (EPA), and the Food and Drug Administration (FDA). According to the framework, the first two groups would evaluate research supported by grants; the USDA, specifically its Animal and Plant Health Inspection Service (APHIS), would regulate genetically altered agricultural plants and animals; the FDA would cover engineered drugs, medical treatments, and foods; and the EPA would handle anything related to pesticides and other chemicals. In that same year, the Toxic Substances Control Act (TSCA) was amended to require a permit from the EPA to release most genetically engineered organisms into the environment, use them in manufacturing, or distribute them commercially for intended release. (Lindow thus had to obtain a permit from the EPA as well as the RAC before carrying out his tests.) This division of labor still applies today.

Most of the products that followed Lindow's lead into open-field testing were plants rather than bacteria. By 1994, more than 2,000 field tests of transgenic plants (those containing genes from another species) had taken place worldwide. No obvious environmental disasters resulted from the tests, and regulation of field testing of genetically altered plants, like regulation of experiments on altered microorganisms before it, slowly relaxed as a result. In 1993, the USDA agreed to require only notification of the agency, not a permit, for testing genetically altered strains of corn, cotton, potatoes, soybeans, tomatoes, and tobacco, which together represented 80 to 90 percent of applications.

On the whole, regulatory agencies in the United States seem to agree with the claims of agricultural biotechnology companies that genetically engineered plants are not substantially different from those changed by traditional breeding techniques and therefore require no additional regulation. Indeed, supporters say genetic engineering is safer than traditional breeding because it allows more precise control of genetic transfer. Critics, however, point out that biotechnologists sometimes combine genes from species so different from one another that they could never be blended by conventional breeding. They believe, therefore, that transgenic organisms should receive more intense scrutiny than those produced by conventional breeding.

Genetically Engineered Plants and Animals

In 1996, when genetically modified crops were first grown commercially in the United States, they covered a mere 4 million acres. By 2003, however, some 96.3 million acres of genetically engineered crops were being cultivated, including up to 80 percent of the country's total soybean crop, 71 percent of its cotton, and 34 percent of its corn. In addition to the United States, countries with large transgenic crop acreage include Canada, Argentina, and China. Some 145 million acres of GM (genetically modified) crops were planted in 13 countries around the world in 2002, according to the Biotechnology Industry Association. The Freedonia Group, a market research organization, estimated that sales of agricultural biotechnology products in the United States alone amounted to more than $2 billion in 2001.

Plants have been engineered to resist insects, disease, drought, salt, and herbicides. Most of these added traits benefit farmers who grow a small number of popular crops, particularly soybeans, corn, and cotton. One of the most widespread types of genetic alteration, for instance, confers resistance to herbicides such as Roundup, Monsanto's brand of the herbicide glyphosate. Agribusiness companies sell seeds for these crops and the herbicide to which they have been made resistant as a package. They often require farmers to sign contracts promising to use only that company's brand of the chemical.

Herbicide-resistant crops let farmers use herbicides more efficiently to control weeds. A farmer who has planted "Roundup Ready" soybeans, for example, needs to spray heavily with that herbicide only once, rather than using other, possibly stronger, compounds repeatedly. The crops therefore reduce the use of herbicides overall, supporters say. The crops also free farmers from the need to plow, reducing loss of topsoil to erosion. These effects help the environment as well as save farmers work and money.

Many environmentalists, however, claim that herbicide-resistant crops provide more threat than benefit. Their chief concern is gene flow, through which genes inserted into crop plants travel to wild relatives. Gene flow is especially easy with a crop like corn (maize), which releases its pollen (male sex cells) into the wind. Wild plants that acquire herbicide-resistance genes, these critics say, could become "superweeds," requiring the use of stronger herbicides that might endanger wild plants on which animals and birds depend for food. Spread of these or other engineered genes could also reduce biological diversity, which is already seriously threatened by the worldwide tendency to plant large areas in genetically identical or near-identical natural crops. This narrow genetic base can turn a plant disease epidemic into a major disaster.

Gene flow may already be occurring. Ignacio Chapela and David Quist of the University of California, Berkeley, claimed in the November 29, 2001, issue of the respected science journal *Nature* that they had found engineered genes in corn from remote fields in Mexico, even though Mexico does not allow planting of genetically altered corn. This was seen as particularly threatening because Mexico is the native home of maize and a center of biodiversity for this crop. Other researchers vigorously attacked the Berkeley scientists' conclusions, even persuading *Nature* to take the unprecedented step of issuing a sort of apology for having printed their paper, but later reports by other scientists seem to confirm it, and many experts in the field feel that gene flow is bound to occur sooner or later. This may not be much of a problem in the United States, where few crop plants have wild relatives growing nearby, but it could be a significant threat in developing countries, where crops and weeds often are related.

Another popular type of genetically altered crop may also cause unexpected environmental damage. These crops have been given a gene from the bacterium *Bacillus thuringensis* (Bt for short) that codes for a natural insecticide. They therefore can produce their own pesticides. (Indeed, the EPA classifies the altered plants as pesticides.) Like herbicide-resistant crops, crops containing Bt have become commonplace on American farms. Farmers who plant these crops can greatly reduce their use of expensive, environment-damaging chemical pesticides, supporters say, but critics have questioned whether the crops really reduce pesticide use significantly.

Organic farmers have used sprays of Bt bacteria as "natural" pesticides for decades, but sunlight destroys the bacteria a few days after spraying. Crop plants containing Bt genes, by contrast, produce the bacterial toxin all the time. This means that pest insects are exposed to it for longer periods and thus have more chance to develop a resistance. Some evidence of increased Bt resistance in pest insects has already appeared. The EPA requires farmers to plant nonengineered crops of the same kind near the engineered ones so that a population of nonresistant pests will be maintained to mate with any resistant ones that may develop, thereby slowing the spread of resistance.

Furthermore, although Bt spray normally does not harm nonpest insects, the same may not be true of the toxin produced by the engineered plants. In May 1999, entomologists at Cornell University published the results of a laboratory study suggesting that pollen from Bt-engineered corn could be carried onto milkweed plants by wind and could then poison monarch butterfly caterpillars feeding on those plants. Later tests under more natural conditions, however, seem to show that milkweed plants do not grow very near corn crops and therefore do not collect enough pollen to harm monarch caterpillars. In any case, biotechnology boosters say, possible harm caused by the pollen must be weighed against the plants' ability to reduce the use of chemical pesticides, which also kill butterflies and other nonpest insects.

Some critics say that genetically modified food crops pose threats to human health as well as to the environment. The first genetically engineered food, a type of tomato called Flavr Savr that had been altered to rot more slowly after ripening and thus be less liable to spoil during shipping and storage, was FDA-approved for sale in 1994. Ever since, groups have demanded the banning, or at least the labeling, of such "unnatural" foodstuffs. However, defenders of the altered foods insist that most are nutritionally no different from their natural cousins and therefore need no special labeling or regulation. The FDA has agreed with this point of view, ruling in May 1992 that genetically altered foods do not need to be reviewed by the agency or labeled as long as they contained no new, toxic, or foreign substances that might cause health problems. Similarly, a summary of 81 scientific studies released by the European Union Commission for Health and Consumer Affairs in 2001 stated that none of the studies found any evidence that GM foods harmed human health or the environment.

The chief health risk from genetically engineered foods probably lies in the possibility that individuals might have allergic reactions (which can be life-threatening) to novel proteins they contain. People allergic to Brazil nuts, for instance, proved also to have allergic reactions to soybeans containing a gene from the nuts. (This problem was detected before marketing, and the nut-containing soybeans were never sold.) Recognizing this potential problem, the FDA requires labels on products to which genes from known food allergens, such as peanuts, have been added.

Warnings of allergies and other health risks from GM food rose dramatically in fall 2000, when environmentalists, beginning with Larry Bohlen of Friends of the Earth in September, found traces of a Bt-containing corn variety called StarLink in numerous corn products used as human foods, ranging from taco shells to corn chips. Although several kinds of Bt corn had been approved for human food, the EPA had approved StarLink only for animal feed because it contained a protein called Cry9C, which, unlike the Bt proteins in the approved types of corn, was not broken down quickly in the human digestive tract. Although there was no evidence that Cry9C actually caused allergies, officials thought that this persistence might allow it to stay in the body long enough for an allergic reaction to develop.

Aventis CropScience, the seller of StarLink corn seed, was supposed to tell farmers to keep it separate from corn intended for food use. Nonetheless, the types of corn somehow became mixed during the storage and milling processes. Eventually more than 430 million bushels of corn were found to be contaminated with StarLink and had to be removed from the food processing chain, and more than 300 types of corn products were recalled.

No one who ate StarLink-tainted corn was ever provably harmed by it. Forty-seven people claimed to have suffered allergic reactions, which were confirmed for 28 of them, but the Centers for Disease Control reported in August 2001 that it had found no antibodies to Cry9C in their blood. It therefore seemed unlikely (though not impossible, the agency admitted) that their immune systems had reacted to the protein. Nonetheless, the incident was an economic and public relations disaster for everyone involved. The recall cost more than $1 billion. Both farmers (including those who had never grown StarLink but had seen the value of their corn crops plummet anyway) and consumers filed class-action suits against Aventis and other companies that had developed and distributed StarLink, and in April 2003 a judge accepted the companies' agreement to pay $110 million to the farmers. (The consumer suit, settled in March 2002, netted only a promise to issue 6 million dollar-off coupons on corn products.) Aventis withdrew StarLink from the market. Opponents of genetically engineered foods claimed that the episode confirmed their worst fears about the difficulty of segregating genetically modified from nonmodified material.

Meanwhile, plenty of people are eating genetically modified material that has been approved for human consumption. Estimates in late 2002 stated that 60 to 70 percent of foods in the United States and Canada contained genetically altered ingredients. No health problems convincingly linked to these foods have been reported, but critics say that this may be because no one has looked for them systematically. Furthermore, since GM foods in North America are not labeled as such, determining whether people with, say, unusual allergies have eaten these foods is almost impossible.

Many environmental and health groups say that regulation of genetically modified crops and foods in the United States is far too lax and too favorable to the biotechnology industry. Numerous recommendations for tightening current regulations have been made, including clarification and streamlining of the division of responsibility among the three agencies involved, making the FDA's premarket reviews of new food crops mandatory instead of voluntary, making reviews open to public input and monitoring, and adding postmarket reviews to detect unexpected health or environmental effects. At the very least, critics say, GM foods should be labeled as such so consumers can decide whether to eat them. Biotechnology spokespeople, however, point out that natural foods can also cause allergies and other health problems, yet they do not undergo the rigorous testing that GM foods receive even under the present system.

Demands for increased regulation of GM foods in America are nothing compared to the outcry against such foods in Europe, where they are often called "Frankenfoods." The European Union (EU) began to require labeling of all foods that might contain genetically modified organisms in 1997, and in 1998 it placed an informal moratorium on approval of new genetically modified crops and on importation of foods that might contain unapproved GM products. Furthermore, on July 2, 2003, the European Parliament approved regulations that would require importing countries to provide traceability of all food products "from farm to fork" and verification that foods not labeled as GM-containing have no more than 0.9 percent accidental GM contamination. EU officials said they would probably lift the GM import ban if the United States agreed to the new regulations, but U.S. farmers nonetheless reacted angrily to the proposed changes, claiming that they were unscientific, unnecessary, and impossible to meet. Calling an earlier draft of the regulations protectionism in disguise, the Bush administration sued the EU in May 2003 for allegedly violating World Trade Organization (WTO) rules. In September, the WTO said it would side with the United States. Nonetheless, the EU rules went into force on November 7, 2003.

This conflict between the United States and Europe has had global effects. The European stance has encouraged other international groups to reject or strictly regulate GM foods. For example, the Cartagena Biosafety Protocol, drafted in Montreal in January 2000 as an addendum to the United Nations Convention on Biological Diversity and signed by more than 130 countries, follows Europe's "precautionary principle" approach to regulation of international shipment of GM crops, assuming the crops to be risky until they are definitely proven otherwise. It requires exporters to obtain permission from importing countries before importing live genetically altered material. This treaty went into effect as international law on September 11, 2003.

Meanwhile, numerous countries have refused to allow importation or planting of GM seeds because they fear that they will not be able to export their crops to Europe if they allow possible sources of GM contamination to enter their supplies. Even the African nations of Zambia, Zimbabwe, Mozambique, and Malawi, which were on the brink of famine, rejected donations of U.S. corn from the United Nations Food Programme in August 2002 because they feared that farmers might plant some of the corn and that, if it proved to be genetically modified, the countries would no longer be able to sell corn to Europe. Some officials also expressed concern about health risks. "We would rather starve than get something toxic," Zambian president Levy Mwanawasa announced.[10]

The African rejection particularly surprised and dismayed biotechnology supporters because they claim that developing nations have even more to gain from the new technology than developed ones. Biotech boosters both outside and within developing countries say that genetically engineered crops can improve food yields and therefore will help feed the countries' poor and hungry populations while, at the same time, preserving the environment by decreasing soil erosion, energy use, and dependence on pesticides and other chemicals. The crops offer other nutritional and medical benefits as well, they maintain. Their favorite example is a crop called golden rice, created in 1999, which contains added daffodil and bacterial genes that allow it to produce beta-carotene, a precursor of vitamin A. Vitamin A deficiency affects between 100 million and 140 million children in poor countries each year and makes about 500,000 of them go blind. Opponents of altered foods, however, say that the effectiveness of golden rice in preventing vitamin A deficiency has been considerably exaggerated. The environmental group Greenpeace, for instance, claims that an adult would have to eat at least 12 times the normal amount of rice in order to obtain the recommended daily dose of vitamin A.

Critics also reject the overall arguments of biotechnology supporters on both sociological and scientific grounds. Widespread hunger, they say, is not caused by lack of food—some countries with large populations of hungry people, such as India, in fact have grain surpluses—but by political, social, and economic problems such as poverty and war, which prevent equitable food distribution. Increased crop yield will bring little benefit unless these problems are corrected, they maintain. Furthermore, they point out, most of the crops that have been genetically engineered so far are ones that well-to-do farmers grow for export rather than ones that poor people grow to feed their families.

Transgenic farm animals have existed since 1985, although the techniques for producing them are still experimental and often fail. Some have been altered to increase their value as food (to produce leaner meat, for example), but many are intended for the branch of the biotechnology industry often called "pharming." They have been given genes to induce production of human hormones, drugs, or other medically useful substances in milk (cattle), eggs (chickens), or even urine (mice). Producing such substances in animals costs less than 10 percent of the amount needed to make them in laboratories or factories, and the substances often can be made more quickly as well.

Pharming is not the only use of genetically altered animals. Mice, for instance, have been given human genes associated with cancer or other diseases, or have had genes of their own made inactive or "knocked out," to make them more useful in experiments aimed at understanding the diseases or in testing drugs. Pigs have been given human genes in the hope that the animals may become compatible donors for organ transplants. Even a primate, a rhesus macaque monkey named ANDi (short for "inserted DNA" backward), was successfully made transgenic in January 2001. ANDi merely contained a jellyfish gene used as a marker, but scientists hope that future transgenic primates, like transgenic mice, will be useful in medical research.

The technology for creating genetically altered animals received a large boost in February 1997, when Ian Wilmut and his coworkers at Scotland's Roslin Institute announced that they had cloned a lamb from an udder cell of a six-year-old Finn Dorset ewe. They named the lamb Dolly, after country-western singer/actress Dolly Parton. The amazing thing about Dolly, from a scientific point of view, was not that she was a clone—a sort of delayed twin of her "mother," with exactly the same genetic makeup—but that she had been made from a mature adult somatic (body) cell, something many scientists had thought could not be done with mammals.

Scientists cloned amphibians in 1952. In 1986 they first cloned mammals with a new technique called nuclear transfer, in which a single cell is fused by electricity with an unfertilized egg from which the nucleus has been removed. The technique was normally applied to embryo cells, but Wilmut and his team found they could use it on a mature cell if they first deprived the cell of nutrients for five days. This starvation put the cell into a resting state, in which it did not divide and most of its genes were turned off. After the cell was fused with an enucleated egg, the cytoplasm of the egg cell somehow reprogrammed the adult cell's nucleus. The combined cell eventually produced a whole animal.

Wilmut's chief aim, and that of many scientists who followed him, was to create a more efficient way to produce herds of identical, genetically altered "pharm" animals. Cloning is potentially better than breeding for producing such animals because with cloning a researcher can be sure that a particular gene or trait existing in the original animal will be preserved, whereas it may or may not be kept in breeding. Cloning may also be a useful way to preserve desirable qualities such as leaner meat or disease resistance in nontransgenic animals.

Within months, Dolly was followed by cloned cattle and other animals, including the first cloned transgenic farm animals (which were also sheep). Most were created with variations of Wilmut's technique. The cloning technique, like techniques for inserting foreign genes into animal germ (reproductive) cells, is still in the early stages of development and very inefficient, however. Dolly, for instance, was the only success out of 277 tries. Many later-cloned animals have died mysteriously before or soon after birth or have been born oversized or deformed. Some studies have suggested that the survivors age more quickly than normal animals. These problems may be due to flaws in the way the genes in the altered egg reprogram themselves. Dolly herself was euthanized in February 2003, at the age of eight years, because she had developed a virus-induced lung tumor. Roslin scientists say this kind of infection is common among older sheep and had nothing to do with Dolly's being a clone, but she may have developed other health problems, such as arthritis, at an unusually early age because of her background.

Nonetheless, some scientists have suggested that cloning could be used to preserve endangered species or even, perhaps, revive extinct ones, such as mammoths. An endangered cattlelike creature called a Javan banteng was cloned in April 2003, the first clone of an endangered animal to survive more than a few days beyond birth. Many conservationists have said, however, that reviving an extinct or nearly extinct species is of little use if the animals' habitat has been destroyed. They believe that money spent on cloning endangered species would be better used for habitat preservation.

So far, transgenic animals have produced much less controversy than transgenic plants, perhaps because these animals have been made in such small numbers. Most complaints have focused on health risks to the animals themselves. For example, animal rights groups have pointed out that a "fast-growing" pig containing growth hormone genes from cows (created by the USDA in the 1980s), was deformed, crippled by arthritis at a young age, and unable to keep itself warm because its body contained so little fat. Such groups also question the ethics of treating animals as mere factories or machines for the production of drugs.

The greatest safety controversy on the animal side of agricultural biotechnology has centered, not on animals that contain altered genes, but on a genetically engineered substance given to normal animals. Beginning in 1985, Monsanto developed a process for making genetically altered bacteria produce bovine growth hormone (BGH, also called bovine somatotropin, or BST). The recombinant hormone (rBGH), trade-named Posilac, is sometimes given to dairy cattle to increase their milk production. It won approval from the FDA in 1993, thereby becoming the first genetically engineered animal hormone approved for use in the United States. No other industrialized country has approved its use.

The FDA has ruled that rBGH is not an additive and therefore does not have to be listed on the label of milk taken from treated cows. A number of groups in the United States and elsewhere have questioned that decision, however. They point out that the hormone is associated with an increased risk of mastitis (inflammation of the udder) and other health problems in treated cows. To prevent or treat the mastitis, dairy farmers give cows antibiotics, some of which can remain in their milk and be drunk by humans. Constant exposure to low doses of antibiotics encourages bacteria that infect humans to become resistant to the drugs. Some critics also worry that insulinlike growth factor 1 (IGF-1), a hormone related to BGH, may be absorbed from the milk of rBGH-treated cows. This substance may increase the risk of breast and prostate cancer.

Fears of rBGH have led to some legal battles. The state of Vermont, suspicious of rBGH from the time the hormone was first tested there in the mid-1980s, passed a law in 1994 requiring that stores label all products that might come from hormone-treated cows. Six industry groups challenged the law in a federal suit, claiming that it interfered with their free speech and with interstate commerce. A U.S. District court in Vermont upheld the law in 1995, but a year later the Court of Appeals for the Second Circuit overruled it. Vermont now has a voluntary labeling law that allows farmers to state that their milk does not come from rBGH-treated cows.

Few other suggestions that transgenic animal products currently present a risk to human health have surfaced so far. The FDA claimed in an October 2003 report that food from cloned animals appeared to present no unusual dangers, for instance. Similarly, a National Research Council report issued in 2002 claimed that most uses of transgenic and cloned animals have little risk of causing allergies or other human health problems. As the time when meat or other products of such animals might enter the food or medical supply approaches, however, calls for strict regulation are increasing. The FDA's Center for Veterinary Medicine announced in October 2002 that it will regulate transgenic animals and their products as "new animal drugs" under the Food, Drug, and Cosmetic Act and therefore will give them the strict review required for all new medicines. In mid-2003, the agency was still working out a policy for regulating cloned animals.

As with transgenic plants, transgenic animals may present a greater threat to the environment than to health. Environmentalists are particularly worried about a transgenic salmon, developed in 2000, which has been given growth hormone genes from a different kind of fish that allow it to grow twice as fast as normal. The creators of the salmon, Aqua Bounty Farms of Massachusetts, say that all their fish will be sterile females, but critics fear that fertile fish might occasionally escape into the wild and then outgrow and outbreed natural salmon, causing an ecological disaster. The National Academy of Sciences' National Research Council shares this concern, according to a report commissioned by the FDA and issued in late 2002.

PATENTING LIFE

Herbert Boyer and Stanley Cohen did not think of patenting their revolutionary gene-splicing technique until their universities insisted that they do so. When they finally applied for the patent in 1974, they donated all proceeds from it to the universities. This altruistic approach did not remain the norm for long, however. Profit motives entered the scene in 1976, when Robert Swanson, a young venture capitalist, persuaded Boyer to join him in founding Genentech (GENetic ENgineering TECHnology), the first biotechnology company based on genetic engineering.

Like many similar businesses that sprang up in the following years, Genentech took advantage of the remarkable reproductive powers of bacteria, the first genetically engineered organisms, and the fact that techniques for growing bacteria in factories already existed. Doubling in number every 20 minutes, bacteria could produce billions of duplicates of themselves—including any genes that had been inserted into them—in just a day or so. All these clones could make the proteins for which the inserted genes coded.

Swanson and Boyer planned to use genetically altered bacteria to make human hormones or other medically useful substances that had been in short supply. They started with insulin, the sugar-controlling hormone that many diabetics must take daily. Insulin from cattle or pigs was easily available, but the animal forms of the hormone differ slightly from the human version, and about 5 percent of diabetics are allergic to them. The product of bacteria containing the human insulin gene, by contrast, was identical to the human hormone and therefore almost never caused an allergic reaction.

When Genentech offered its stock to the public for the first time in October 1980, it had produced genetically engineered human insulin experimentally (starting in 1978) but had not yet won federal permission to sell it. Nonetheless, belief in the company's future was so strong that the price of its stock jumped from $35 to $89 a share in the first few minutes of trading. By the time Genentech finally won FDA approval to sell its genetically engineered insulin in 1982, companies that had followed its lead were using modified bacteria to produce some 48 different hormones or other substances from the human body, as well as several drugs and vaccines. The business side of genetic engineering was well under way.

Where money goes, lawyers and lawsuits soon follow, and biotechnology has been no exception. Most of the suits and other legal disputes in the industry have concerned patents and other ways to protect intellectual property. Genetic engineering has broken new ground in the patent field by raising the question of when—or, indeed, whether—living things, body parts, cells, genes, or DNA sequences should be patented. Is it possible or right to "own" a piece of life?

Patents are a time-honored way of encouraging and rewarding inventiveness by giving an inventor the exclusive rights to an invention's use and sale. The doges of Venice and the rulers of England, among others, traditionally granted monopolies, or rights of exclusive sale, to inventors. The term *patent* came from the "letters patent," or open letters addressed to the general public, that kings used to proclaim these monopolies.

Article 1, section 8 of the U.S. Constitution grants Congress the right to "promote the progress of science and useful arts, by securing for limited times to authors and inventors the exclusive right to their respective writings and discoveries." Patents are issued by the Patent and Trademark Office (PTO), which is part of the Department of Commerce. In 1793, Congress defined a patentable invention as "any new and useful art, machine, manufacture or composition of matter." The Patent Act of 1952 changed the word *art* to *process* and added the criterion that, as well as being new and useful, an invention must not be obvious "to a person of ordinary skill in the art."[11]

Traditionally, products of nature, including living things, were not held to be patentable because they were not inventions, or objects created by human ingenuity. An exception of sorts was made for plant varieties, which were protected in the United States by the Plant Patent Act (1930) and the Plant Variety Protection Act (1970). These laws allowed breeders who had developed new varieties to block others from reproducing them.

Genetic engineering brought a major change to the U.S. patent system in 1980, when the Supreme Court ruled by a 5-to-4 vote that a patent could be issued to Ananda Chakrabarty, a scientist working for General Electric Corporation, for a genetically altered bacterium that digested petroleum and could be used to help clean up oil spills. In his majority opinion, Chief Justice Warren Burger quoted Congressman P. J. Federico as saying that Congress had intended the 1952 patent law to apply to "anything under the sun that is made by man"—which, Burger decided, could include living things, if humans had altered them.[12] Chakrabarty's patent was the first patent issued for a living organism.

The *Diamond v. Chakrabarty* ruling was soon extended to more complex organisms. Plants, seeds, and plant tissue cultures became patentable in October 1985, when Molecular Genetics, Inc., obtained a patent on a type of genetically engineered corn. A year and a half later, in April 1987, the PTO ruled that genetically engineered animals (except humans), as well as human genes, cells, and organs, could also be patented. The first genetically engineered animal to be patented was the Harvard Oncomouse, a type of mouse designed to be a test animal in cancer research. It was patented in April 1988 by Harvard University. The PTO received 1,502 patent applications for transgenic animals in the decade that followed and approved more than 90.

Most were genetically altered mice intended for medical research, but the list ranged from worms to sheep.

Not everyone has agreed that patenting living things is ethical. In 2000, for example, the Council for Responsible Genetics drafted a "Genetic Bill of Rights," which states that "all people have the right to a world in which living organisms cannot be patented."[13] The European Union's Directive 98/44, on the legal protection of biotechnological inventions, states that patents may be denied or restricted if they are considered to violate public morality, and groups that oppose genetic engineering and patents on living things, such as the European Green Party, have frequently (although not always successfully) called on this feature of the directive when protesting particular biotechnology patents.

Other criticisms of the biotechnology industry's stress on patents, including those on living things, have come from scientists. They point out that, in the United States at least, an invention cannot be patented if details of it have been published more than a year before the patent application is filed. Researchers thus often withhold data until applications can be submitted, impeding the free flow of information that has been traditional in science. This complaint is ironic, considering that patents—which require a description of an invention detailed enough to allow anyone skilled in its field to make the device—were intended to increase access to information about technology. Furthermore, as more and more scientists, including those whose main work is done at universities, become involved in biotechnology ventures, the urge to patent may affect the objectivity of their work. At the very least, such conflicts affect others' perception of the scientists' objectivity. For example, the objectivity of a scientist's study of the effectiveness of a drug might be questioned if the scientist was part owner of a patent on the drug or had stock in a company that manufactured it.

Patents also create monopolies that can reduce the supply of products, increase their price, and limit competition. This is potentially a particular problem in agricultural biotechnology, where giant companies such as Monsanto and DuPont often control both genetically engineered crop seeds and the chemicals on which they depend. Some farmers, especially in the developing world, may not be able to afford these products or to compete against those who can. Furthermore, critics of agricultural biotechnology patents say, such patents force farmers to buy new seeds from the companies every year at prices they can ill afford, rather than planting seeds from the previous year's crop as they traditionally do. Some companies are even developing genetic techniques that automatically enforce this limit by making plants that cannot reproduce. "Patents create a monopoly that threatens food security and the livelihood of farmers," says Anuradha Mittal, cofounder of the Institute for Food and Development Policy in California.[15] Biotechnology companies, however, point out that on occasion they have donated patented products to poorer countries at reduced or no cost.

The question of patents on living things becomes even more politically charged when the patents are for altered—or even, sometimes, unaltered—creatures that originally came from the poverty-stricken Tropics. Critics such as the ETC Group (Action Group on Environment Technology and Concentration, formerly Rural Advancement Foundation International) claim that large companies perform acts of "biopiracy," raiding the developing world—the source of most of the planet's genetic diversity and home to most of its undiscovered species—for specimens or genetic samples that may prove useful. These companies, the critical groups say, provide little or nothing in return to the land that nurtured the plants or the native farmers and healers who used or developed them. They may even sell finished drugs or other products made from these plants back at high prices to the countries that originated them.

A widely discussed example of alleged biopiracy involved the neem tree, which the people of India have used to fight bacteria, fungi, and pest insects for millennia. In 1994, the USDA and W. R. Grace, a large multinational corporation, obtained a European patent for using neem oil against fungi that infest plants. Indian farmers, fearing that the patent would keep them from using their traditional remedies, raised massive protests. (Biotechnology spokespeople claim that the patent would not have affected practices in existence before it was granted.) The European Patent Office revoked the patent on May 10, 2000, and crusaders against such patents, such as Indian activist Vandana Shiva, claimed a victory.

Protests have also arisen over exceptionally broad patents, beginning with the "process" and "product" patents granted to Boyer and Cohen, which seemed to cover all of recombinant DNA technology. A later dispute of this kind related to patents granted in the early 1990s to Agracetus, a subsidiary of W. R. Grace, for all genetically engineered cotton and soybeans, regardless of the method by which they were produced. "It was as if the inventor of the assembly line had won property rights to all mass-produced goods," complained Jerry Caulder, chief executive officer (CEO) of a rival company, Mycogen.[14] Broad patents, particularly on basic processes or materials necessary for a whole field of research, can make research in that field so expensive that small companies cannot take part in it.

A related subject of dispute is so-called reach-through rights, which give the owner of a patented technology royalties on all commercial products created by that technology, regardless of who develops them. Companies often try to obtain reach-through rights from licensees as a condition of licensing their technologies. Complaints from other companies or, sometimes, government bodies such as NIH have caused the PTO to reverse its decisions on some broad patents, such as one of those granted to Agracetus, and have forced some patent-holding companies to moderate their demands for reach-through rights.

In answer to these complaints, biotechnology industry spokespeople insist that patents on genetically altered living things and the processes that produce them are both ethical and necessary. Patents, they note, traditionally have been held to benefit society by encouraging invention and giving inventors an inducement to risk the time and money necessary to bring inventions to the marketplace. Patents also forced inventors to make public the details of their technology, thus (once the patent expired) encouraging business expansion and competition. Patents, supporters claim, are necessary to give modern biotechnology investors an incentive to support the research and testing needed to develop, say, a new drug. Such a development process can take seven to 10 years and cost up to $500 million. Biotech boosters claim that society ultimately benefits from industry patents because patented research can produce such things as increases in world food supply and life-saving medicines.

As for "biopiracy"—or "bioprospecting," as its supporters prefer to call it—both sides of the debate cite as an example an agreement made in September 1991 between the National Biodiversity Institute (INBio) in Costa Rica and American drug giant Merck & Co. INBio promised to provide samples of Costa Rican wild plants, microbes, and insects—indeed, what foes say is the entire genetic wealth of the country—to Merck to screen for possibly useful drugs. In return, Merck offered to give INBio a research and sampling budget of $1.135 million, royalties from any commercial products that result from the samples, and training and assistance to help Costa Rica set up its own drug screening program. INBio promised to give 10 percent of its initial payment and 50 percent of its royalties to a fund used to conserve the country's national parks. Supporters say that this sort of agreement provides excellent incentives for countries to preserve their biological diversity, whereas opponents see it as selling a country's genetic birthright. Several international agreements, such as the Convention on Biological Diversity and Intellectual Property Rights, signed by 157 nations (not including the United States) at the Earth Summit in Rio de Janeiro in June 1992, have included sections that attempt to protect countries' biological and genetic resources.

As with labeling of genetically engineered foods, differences in public feeling and law between the United States and other parts of the world have caused conflicts in regard to patenting living things. For instance, in 1998 the European Union Parliament and Council decreed in Directive 98/44 that processes for genetically modifying animals that might cause them suffering cannot be patented unless "substantial medical benefit" is likely to be derived from them, although altered plants and processes for making them can be patented. The Supreme Court of Canada decreed in December 2002 that "higher life-forms," defined as multicellular differentiated organisms including both plants and animals, cannot be patented under that country's current patent law. On the other hand, Japan, like the United States, permits patenting of higher life-forms.

The United States has repeatedly pressured other nations to strengthen their legal protection of intellectual property in biotechnology, including patented living things. In 1995, largely at U.S. instigation, the World Trade Organization (WTO) formulated a controversial agreement called the Trade Related Intellectual Property Rights Agreement (TRIPS), which requires WTO members to make their national patent laws similar those in the United States. The supposed aim of TRIPS is to reduce impediments to international trade so that all countries can obtain the benefits of biotechnology (companies will not deal with countries that do not offer them patent protection), but opponents say that the agreement's real purpose is to guarantee protection and high profits to multinational companies. Many developing countries have expressed resentment of TRIPS, saying that installing a U.S.-type patent system would be prohibitively expensive and would bring little, if any, benefit to them.

Types of human cells and natural human body chemicals have also been patented in the United States and Europe by companies that developed processes for isolating them, even if the cells or chemicals themselves were not changed. When a line of laboratory cells can be traced to the body of a particular individual, however, disputes have arisen about who owns the rights to it.

The best known case of this kind in the United States began in 1984, when Seattle businessman John Moore sued his doctor, the University of California, and several drug companies. Moore's suit claimed that the drug companies had made and patented a profitable laboratory cell line from his spleen, which had been surgically removed as a cancer treatment in 1976. Moore claimed that, although he had consented to the operation, he had not been told about the commercial use of his spleen cells, which he said had been planned even before the surgery. He maintained that the cells were still his property even after they had been removed from his body and that they should not have been used without his permission.

A lower court supported Moore, but in 1990 the California Supreme Court ruled against him, stating that patients do not have property rights to their tissues. The court's majority opinion said that granting such rights would "destroy the economic incentive to conduct important medical research."[16] Not all states have followed California's policy, however. In July 1997, for instance, Oregon passed a law granting ownership of tissue and all information derived from it to the person from whom the tissue came.

In the 1990s the cutting edge of biotechnology patenting moved from organisms and cells to pieces of the genetic code itself. Aided by such inventions as the automatic gene sequencer and the polymerase chain reaction (PCR), which allows tiny pieces of DNA to be rapidly copied many times for analysis, scientists in the late 1980s began to work out the base sequences of the genomes, or complete collections of genes, of living things. The most ambitious genome sequencing project of all has been the Human Genome Project, launched in 1990. Even as that project was being completed, the government and private groups involved in it were arguing about who would have rights to the results. They eventually agreed that the raw sequence data would be published on the Internet, where anyone could view it.

The U.S. Patent and Trademark Office (PTO) has permitted the patenting of unaltered genes or DNA sequences, including human ones, since 1987, provided that they have been isolated and purified and that they fulfill the standard patent requirements of novelty, usefulness, and unobviousness. There has been considerable argument, however, about the degree of usefulness that an applicant for a gene or sequence patent must demonstrate. PTO guidelines issued on January 5, 2001, state that applicants must demonstrate "specific, substantial, and credible" utility. By early 2002, the PTO had granted about 1,300 patents on human genes. Europe also permits patenting of unaltered human genes, but the European Court of Justice ruled in 2001 that this allows "only inventions combining a natural element with a technical process making it possible to isolate that element or reproduce it with a view to an industrial application" to be patented, which means that genes and gene sequences can be patented only for particular applications.[17]

Critics of gene and DNA sequence patenting question whether isolation and purification should be considered sufficient to qualify naturally occurring genes or gene segments as "inventions." They also claim that gene patents tie up research information, including information obtained with public funds, and block the development of tests or cures for diseases. Some patient groups, too, fear that patents on genes such as BRCA1 and BRCA2, involved in human breast cancer, will make diagnostic tests using these genes too expensive for most people to afford.

This has already happened, according to a lawsuit filed by several families with a rare genetic disorder called Canavan disease. They donated samples of blood and tissue to Miami Children's Hospital, which used the material to identify the gene that causes the disease in 1993. The hospital patented the method for detecting the gene and then, the families say, adopted such a restrictive licensing policy that a diagnostic test using the gene became too expensive for others who might have the disorder to afford. The families sued for breach of contract and violation of informed consent requirements, stating that the hospital had implied that if they donated their blood, any resulting test would be made available to everyone who needed it. The suit was filed in October 2000 and, in late 2003, was still pending.

To prevent this kind of problem, a group of families with a second rare inherited disease of the skin and blood vessels, pseudoxanthoma elasticum, has insisted on sharing control of any patent on "their" gene with the University of Hawaii as a condition of donating biological material to the university. In spring 2001, the university tentatively agreed to the family organization's demand to have a say in licensing decisions and receive 50 percent of royalties resulting from a gene patent.

DNA "Fingerprinting"

DNA "fingerprinting" has been used to identify lost children and straying fathers, reveal the genetic makeup of endangered species, and trace the origins of poached elephant tusks. It proved that bones found in Siberia belonged to the last Russian czar and disproved the claims of a woman who had insisted for decades that she was his surviving daughter. More important to most people, it has convicted—or exonerated—hundreds of people accused of rape and murder. DNA identification is the most direct application of biotechnology to the courtroom.

DNA profiling, or identification testing, was invented by Alec Jeffreys, a geneticist at the University of Leicester in Britain, in the early 1980s. In 1985 Jeffreys published a paper describing it in the journal *Nature.* He based his test on the fact that certain regions in human DNA consist of short base sequences repeated over and over. These regions are not the protein-coding part of genes (exons) but rather so-called junk DNA (introns), the biological purpose of which (if any) is still unknown. Jeffreys first used regions called minisatellites, but he soon changed his focus to other stretches of DNA called variable numbers of tandem repeats (VNTRs). Unlike most genes, which exist in only a few different forms, VNTRs vary considerably from person to person, even within the same family, in the number of repeated sequences they possess. Jeffreys concluded that the chance that two people, other than identical twins (who have exactly the same genetic makeup), would have the same numbers of repeats at, say, five different spots was vanishingly small. Comparison of DNA from a blob of semen or a bloodstain left at a crime scene with the blood of a suspect, therefore, should be able to show with high accuracy whether the crime evidence came from the suspect.

Jeffreys's DNA profiling technique was first used to verify family relationships in immigration and paternity disputes. It entered the criminal court in 1987, when, investigating the rape and murder of two teenaged girls, police in Leicestershire took the unprecedented step of asking all men between the ages 13 and 30 years in three villages near the crime scenes—about 5,000 people—to give blood samples for comparison with semen found on the bodies of the girls. At the officers' request, Jeffreys agreed to analyze all the specimens that could not be quickly eliminated on the basis of blood type, some 40 percent of the total.

While Jeffreys was testing, police received word of a conversation in a local pub in which a man had bragged about having given a blood sample in a coworker's stead. They questioned the man, and he identified the reluctant coworker as a 27-year-old baker named Colin Pitchfork. Faced with evidence of his falsification, Pitchfork confessed to the crimes, and when his blood was finally tested, its DNA matched that in the semen samples. Pitchfork went to trial in September 1987, was found guilty, and was sentenced to life in prison, thus becoming the first criminal convicted by DNA evidence. DNA testing showed its other side—its ability to clear innocent people—in the same case by demonstrating that 17-year-old Rodney Buckland, who had confessed to one of the murders, could not have committed it because the DNA in his blood did not match that in the semen samples.

A mere month after Pitchfork's trial, DNA evidence was used in a rape case in the United States as well (*Florida v. Andrews*). There, too, it eventually led to a conviction. Police and prosecutors began to regard DNA testing as the greatest aid to criminal identification since the introduction of fingerprinting a century before. Judges and juries, for their part, were awed to hear experts testify that the odds of a suspect's DNA matching that in a crime scene sample by chance were, say, one in a trillion. As one juror in an early DNA case said, "You can't argue with science."[18]

Within a year or two, however, people did begin to argue with science. No one, then or since, disagreed with the basic principles behind DNA testing, but lawyers and scientists questioned both the accuracy with which the tests were carried out and evaluated and the methods used to derive the statistics that sounded so impressive in a courtroom.

From DNA profiling's first uses to the present day, the accuracy with which the tests are performed has been questioned in certain cases. Sometimes accuracy is fatally compromised by the mishandling of samples before they even reach the testing laboratories, as the defense suggested so successfully in the famous O. J. Simpson murder trial in 1995. Testing laboratories and testing kits also vary in quality and, until 1994, despite repeated requests from scientists in the field, there was no federal program for licensing or giving proficiency tests to laboratories that performed DNA analysis. In that year, Congress passed the DNA Identification Act, which gave responsibility for training, funding, and testing DNA profiling laboratories to the Federal Bureau of Investigation (FBI). The FBI has established quality assurance standards and an accreditation program for forensic DNA testing laboratories, but it does not test the laboratories directly. In 1997, furthermore, the agency's own DNA testing laboratory was shown to have, as Deputy Attorney General Jamie Gorelick admitted, "a serious set of problems" in the accuracy of its work.[19] FBI spokespeople say that the laboratory subsequently improved, but a CBS online report in April 2003 stated that the Justice Department inspector general, an independent watchdog, was examining the laboratory again in the wake of the discovery that a technician there had failed to follow proper DNA testing procedure for two years before being detected.

In the early 1990s, some scientists and lawyers also questioned the statistics used in court to show the probability of an accidental match between suspect and crime scene samples. To begin with, as critics have pointed out, the frequently used term "DNA fingerprinting" (coined and trademarked by Jeffreys) is somewhat misleading. Each person's fingerprint is unique, and so (unless the person has an identical twin) is each person's DNA. DNA identification tests, however, do not examine a person's whole genome, but only tiny fragments of it. The chance that someone else will have the same sequence in those fragments is usually very small, but, it is not zero, especially if two suspects belong to the same family or small subset of an ethnic group. Thus, although lack of a match in a DNA test could conclusively prove innocence (assuming that the test was properly carried out), critics maintained that a match did not conclusively prove guilt.

In the early 2000s, although mandatory national standards for DNA testing laboratories still have not been established, the overall accuracy of DNA profiling has improved enough that it is seldom questioned in court cases if law enforcement personnel and laboratories have followed proper procedures for evidence preservation and testing. Modern testing is usually automated, which reduces the risk of error. The standard DNA identification test used in the United States today checks sequences called short tandem repeats (STRs) at 13 different spots in the genome, or loci, where the sequences are known to vary considerably. The chances that two unrelated people have the same DNA patterns at all 13 of the tested loci are said to be less than one in 100 billion—perhaps as low as one in a quadrillion. For all practical purposes, DNA identification has become the unique "fingerprint" that Alec Jeffreys claimed. Furthermore, unlike the earliest DNA tests, which could take several months and required relatively large samples in good condition, today's tests can be performed in hours, can use degraded samples, and can obtain accurate identification from a bit of saliva, a single hair root, or even a few skin cells left behind by a person typing on a computer or clutching a cell phone.

Most identification testing uses DNA from the cell nucleus, but in some cases the test is done on DNA from the mitochondria, tiny bodies that control the cell's use of energy. Unlike nuclear DNA, which is inherited from both parents, mitochondrial DNA comes only from the mother. Thus, relatives closely related on the female side of a family—brothers and sisters, for instance, or grandmothers on the mother's side and their grandchildren—will have the same or nearly the same mitochondrial DNA. Because of this, identification through mitochondrial DNA is not as precise as that given through nuclear DNA. However, the fact that living relatives on the mother's side can provide matching samples makes mitochondrial DNA testing particularly useful in resolving questions about family relationships and tracing missing persons. Mitochondrial DNA has been used, for instance, to identify children abducted during wars so they could be reunited with their families. The FBI forensics laboratory has created the National Missing Persons DNA Database, which relies on mitochondrial DNA from relatives of the missing people.

In addition to being less precise, testing mitochondrial DNA is considerably more difficult and expensive than testing nuclear DNA. Nonetheless, tests on mitochondrial DNA can prove helpful when nuclear DNA is not available. For example, mitochondrial DNA is present in hair, bones, fingernails, and teeth, which lack nuclear DNA, and it is also much more resistant than nuclear DNA to degradation by time and environmental factors such as sunlight. It thus can be used to identify badly decomposed bodies or skeletal fragments.

Today the chief subject of debate related to DNA testing is the preservation of samples or identification profiles in large databases. Britain established the first national database for DNA profiles of convicted felons, the Criminal Justice DNA Database, in 1995. It was to contain profiles of everyone convicted of a crime serious enough to warrant imprisonment. Germany, France, South Africa, Canada, China, Australia, and a number of other countries now also have databases of DNA taken from convicts.

In the United States, all 50 states have authorized DNA collection from at least some groups of convicted criminals and establishment of state databases, although not all the databases are actually functional. At the federal level, the FBI opened its National DNA Index System (NDIS) in 1998. The bureau also manages the Combined DNA Index System (CODIS), software that coordinates the national, state, and local databases so that police can match samples at crime scenes with DNA profiles of convicted felons anywhere in the country. In late 2002, *Time* magazine claimed that users of CODIS had access to more than 1.2 million DNA profiles.

Both prosecutors and defense attorneys have found DNA testing and DNA databases invaluable. The technology has allowed police to solve "cold cases," in which crimes were committed years or even decades before. In California, for instance, DNA testing led to the conviction of Larry Graham in August 2002 for raping and murdering a five-year-old girl, Angela Bugay, in 1983. DNA testing has also conclusively proven the innocence of prisoners held for decades, resulting in their release. By mid-2003 the Innocence Project, established in 1991 by New York City lawyers Barry Scheck and Peter Neufeld to obtain DNA testing for convicted felons who declared their innocence, had freed 131 people, some of whom had been on death row. Such results have made some states place a moratorium on enforcing death sentences. In cases such as *Harvey v. Horan* (2001, 2002), some convicts have even asserted a constitutional due process right to postconviction DNA testing, but as of mid-2003, no higher court has agreed with them.

Because DNA databases have proven so useful, many states and countries have expanded their lists of crimes for which DNA samples are required from convicts. Some now take samples from people convicted of all felonies, and a few include certain misdemeanors as well. Proposals have even been made to take samples from all people who are arrested. Such suggestions are unlikely to be acted on in the near future because state and national databases already possess hundreds of thousands of samples that they lack money and facilities to test. If promised increases in federal funding and improvements in technology reduce this backlog, however, the idea of expanding criminal databases may become more popular.

A few supporters of DNA databases, such as Akhil Reed Amar, a professor at Yale Law School, have even proposed that national databases be created to contain genetic records from all citizens, perhaps obtained at birth. Britain already seems to be moving in this direction. In March 2003, the British government changed the name of its database to the National DNA Database and announced that police would have the right to keep indefinitely DNA identity information for anyone from whom samples were taken, even if the people were never charged with a crime. At that time the British Forensic Science Service was said to have DNA profiles of more than 2 million people and to be adding about 2,000 samples a week. The U.S. Department of Defense also has a noncriminal DNA database, which it says it keeps to identify the remains of soldiers killed in action. In addition, numerous private laboratories have "biorepositories" with hundreds of thousands of blood or tissue samples taken for genetic testing, for instance to establish paternity, and then stored.

Groups such as the American Civil Liberties Union (ACLU) and the Libertarian Party see DNA databases as a potentially serious threat to civil liberties, particularly if they include people who have not been convicted of crimes or consist of actual tissue samples rather than just identification profiles. They say that the taking of blood samples, at least, requires an intrusion into the body that may violate Fourth Amendment protection against unreasonable searches and seizures. In addition, although DNA identification itself does not reveal personal health information because the tested loci are not actually genes, preserved blood or tissue samples would contain such information, which might lead to discrimination if it fell into the hands of employers or insurers.

So far, courts have found the taking of samples from convicted and imprisoned felons, at least, to be constitutional. For instance, in 1997 a group of female death row inmates in California challenged the constitutionality of a program that required convicts to donate their DNA, claiming that it violated their Fourth Amendment rights and their right to privacy. A Sacramento Superior Court judge granted them an injunction in 1998, saying that the program was constitutional but the sampling procedure was flawed. In May 2002, however, the Third District Court of Appeals unanimously rejected their suit *(Alfaro v. Terhune)*, and the California Supreme Court refused to review the case, letting the appeals court decision stand. Furthermore, California governor Gray Davis signed legislation in September 2002 that allows prison personnel to take such samples through the use of "reasonable force" if necessary. In October 2003, however, a panel of the Ninth U.S. Circuit Court of Appeals struck down a federal law requiring parolees not suspected of new crimes to give blood samples for a DNA databank. Bowing to a Bush administration request, the full court has agreed to reconsider the ruling.

The courts are likely to feel even more strongly about people not convicted, or perhaps not even accused, of any crime. Useful as an all-citizen database might be in capturing criminals, tracing missing persons, and perhaps in preventing crime, many people feel that it would not be worth the price. "The inherent danger to our conception of ourselves as a free and autonomous society requires that further expansion of the preventive state, represented by the creation of a universal [DNA identification] database, be vigorously opposed," writes Rebecca Sasser Peterson of the Georgetown University Law Center in *American Criminal Law Review*.[20]

GENETIC HEALTH TESTING AND DISCRIMINATION

Leroy Hood, a pioneer in the development of automated genome analysis machines, said in 1987 that in the next century, "When a baby is born, we'll 'read out' his genetic code, and there'll be a book of things he'll have to watch for."[21] Completion of the Human Genome Project and development of technology such as DNA chips (sometimes called gene chips or biochips), devices that use DNA on a computer chip to scan thousands of genes at a time, are bringing the future described in Hood's prediction rapidly closer. Some people wonder, however, whether such detailed knowledge of each individual's genes will be an entirely good thing. On the one hand, it could help people avoid illnesses to which their genes predispose them by, say, making certain lifestyle choices or undergoing tests for certain conditions unusually often. It could help physicians tailor drugs and other medical treatments precisely to the needs of individual patients. On the other hand, critics say, it could also lead to a new form of discrimination, based not on race or gender but on one's genes. "Genetic discrimination is the civil-rights issue of the 21st century," says Martha Volner, health policy director for the Alliance of Genetic Support Groups, an organization for families with inherited diseases.[22]

Potentially, as the ability to analyze individual genomes increases, genetic discrimination could affect anyone. "All of us have something or other in our genes that's going to get us in trouble," says Nancy Wexler of Columbia University, a leader in research on the form of inherited brain degeneration called Huntington's disease—and at risk for the disease herself because her mother had it. "We'll all be uninsurable."[23]

Genetic discrimination is most likely to limit access to life and health insurance, the latter of which, at least in the United States, is more or less required for quality health care. It could also bar people from jobs because most large employers provide health insurance for their workers, and employers may not want to risk having their group insurance premiums raised by hiring workers who seem likely to become sick. Genetic discrimination might even affect marriages and family relationships.

Discrimination on the basis of genetic makeup, as it was shown in characteristics thought to be inherited, existed before the word *gene* even entered the English language. This discrimination arose from the so-called science of eugenics (the word means "well born"), which was founded in the late 19th century by Francis Galton, a cousin of Charles Darwin. Galton believed that intelligence and other personality characteristics were inherited. He said that people with desirable characteristics (as he and his social group defined them) should be encouraged to have children, whereas people with undesirable characteristics (often meaning characteristics of other ethnic groups or social classes) should be discouraged or even forcibly prevented from doing so. In 1872 he wrote:

> *It may become to be avowed as a paramount duty, to anticipate the slow and stubborn process of natural selection, by endeavouring to breed out feeble constitutions, and petty and ignoble instincts, and to breed in those which are vigorous and noble and social.*[24]

Galton and his followers saw such practices as the human counterpart of scientific animal and plant breeding.

Around the start of the 20th century, Galton established the Eugenics Society in Britain to carry out his aims. Similar groups were formed in the United States and Germany and attracted many members. People in the American and European middle and upper classes saw eugenics as an answer to both the (perceived) prolific breeding of the lower classes, especially immigrants, and the economic costs of caring for the physically and mentally disabled. Eugenics supporters in the early years of the century included both respected scientists and such well-known nonscientific figures as Theodore Roosevelt, George Bernard Shaw, and (in his youth) Winston Churchill.

Negative eugenics—forcible blocking of reproduction for those thought not fit to reproduce—was codified into law in many places. By the 1930s, some 34 states in the United States had passed laws requiring the forcible sterilization of criminals, the mentally retarded (developmentally disabled), the insane, or others considered unfit to reproduce. According to one estimate, about 60,000 Americans were sterilized as a result of these laws.

In the landmark 1927 case of *Buck v. Bell,* even the U.S. Supreme Court upheld the forced sterilization of a developmentally disabled woman, Carrie Buck. Noting that both Buck's mother and her seven-month-old daughter also seemed to have subnormal intelligence, renowned justice Oliver Wendell Holmes wrote in his majority opinion on the case, "Three generations of imbeciles are enough."[25] (In fact, scientist-writer Stephen Jay Gould reported in a famous 1984 essay that Carrie Buck had been reexamined in 1980 and was found to be of normal intelligence, and school records suggested that her daughter had been normal, too. The only "deficiencies" of the three generations of Bucks, Gould concluded, were that they were poor, uneducated, and violated contemporary sexual mores by giving birth to children out of wedlock.)

Denmark, Finland, Norway, Sweden, and some Canadian provinces also enacted eugenics laws in the 1920s or 1930s. Germany passed a eugenic sterilization law in 1933, soon after the Nazis took power, and used it to force sterilization of some 400,000 institutionalized people whose defects were presumed to be inherited. Later the Nazis carried eugenics to its extreme by trying to wipe out the genes of whole ethnic groups through mass killing.

Partly because of the Nazi excesses, the tide of public opinion began to turn against eugenics around the time of World War II. In *Skinner v. Oklahoma* in 1942, for instance, the Supreme Court struck down a state law requiring forced sterilization of convicted criminals, saying that procreation is "one of the basic civil rights of man."[26] Belief in eugenics never totally died out, however. Many eugenics laws remained on the books until the 1970s, although they were seldom enforced. As late as 1980, 44 percent of respondents in a U.S. poll favored compulsory sterilization of habitual criminals and the incurable mentally ill. Even today, China has a law that can deny permission for marriage to or force sterilization or abortions on people with certain inherited diseases.

In the 1970s, just as traditional eugenics was fading, a new form of genetic discrimination began to appear. Tests that examine DNA directly were still decades away, but sufferers from and carriers of certain inherited diseases could be identified indirectly by tests for particular substances in their blood or other body fluids. Chief among these illnesses was sickle-cell disease (also called sickle-cell anemia), a blood disease that affects mostly people of African ancestry.

Sickle-cell disease is caused by a single recessive gene that codes for an abnormal form of hemoglobin, the protein in red blood cells that gives them their color and allows them to carry oxygen through the body. Cells containing the abnormal hemoglobin take on a crescent shape and block the body's smallest blood vessels, depriving tissues of oxygen and causing considerable pain, disability, and sometimes death. Only people who inherit defective genes from both parents suffer from the disease, however. Those who inherit a defective gene from one parent and a normal one from the other—called sickle-cell carriers or possessors of the sickle-cell trait—can pass the disease to their children, but they themselves are quite healthy. Indeed, they may have had a health advantage in Africa, because sickle cell trait seems to be associated with resistance to malaria, an endemic disease in much of the continent. About one in 500 African Americans suffers from sickle-cell disease, and one in 10 is a carrier.

Tests on hemoglobin in blood samples can detect sickle-cell disease and sickle-cell trait, and in the early 1970s African Americans were widely screened for this condition. The screening programs were intended to identify carriers so they could be counseled about the risk of having children with the disease if they married other carriers. Because of widespread lack of understanding about the difference between sickle-cell carriers and people with the full-blown disease, however, some carriers were charged high insurance premiums or denied certain kinds of employment. The U.S. Air Force and some private commercial airlines, for instance, barred sickle-cell carriers from being pilots because of an unproven belief that such people would have trouble with the decreased oxygen levels at high altitudes.

African Americans supported sickle-cell screening programs at first, but they soon began to protest them as racism in a new, medical guise. A number of states therefore either dropped the programs or passed laws limiting use of information obtained from them. North Carolina, for instance, passed a law in 1975 that prohibited employers from discriminating against anyone possessing sickle-cell trait or the type of hemoglobin associated with it. This was the first American law to address genetic discrimination in the workplace.

The same issues raised by sickle-cell screening in the 1970s reappeared in the 1990s as a growing number of tests became available to screen DNA for mutations that cause or increase the risk of particular diseases. By early 1999, genetic centers could test for 30 to 40 inheritable conditions, in some cases with an accuracy of 99 percent. These included not only classic inherited diseases such as Huntington's but some forms of more common and complex illnesses, such as cancer and heart disease, which can have a genetic component.

In the early 2000s, all U.S. states had established programs to screen newborn babies for certain inherited diseases, ranging from three or four to more than 30 conditions. (These tests are not, strictly speaking, genetic tests; they reveal inherited conditions by identifying certain biochemicals in the blood.) Most of the conditions tested for are ones for which early intervention, for instance in the form of diet restrictions, is essential to prevent severe damage, and bioethicists generally feel that screening of babies should be limited to such diseases. Other genetic tests, they say, should be given only to adults who can consent to them. Although the screening tests can be lifesaving to a few children, some ethicists and civil libertarians worry that private genetic information could later be obtained without consent from the blood spots on the resulting "neonatal cards," which hospitals usually keep after testing.

Critics have pointed out a number of problems with the current generation of genetic tests. First, the test results are often hard to interpret and even harder to explain to tested people and their families. Tests and testing laboratories are also poorly regulated, and results may not be accurate. In many cases, too, the tests do little to improve health because people and their physicians cannot prevent or cure the diseases that the tests predict. Finally, the tests make people focus on individual genes as a cause of disease when, in fact, most illnesses—including the most common and deadly ones, such as cancer and heart disease—result from the interaction of multiple genes with each other and with environmental factors such as lifestyle choices (smoking, eating a high-fat diet) or exposure to toxic chemicals. Thus, even if a person inherits a gene that increases the risk of, say, heart disease, that person may never develop the disease, whereas someone else with a better genetic profile but a worse assortment of environmental exposures may.

Additional ethical problems arise because the results of genetic tests affect not only individuals but their families. Suppose, for instance, that a woman learns from a test that her son will develop Duchenne muscular dystrophy, an inherited condition passed from mothers (who do not show signs of illness) to sons that causes disability and early death. The woman's sister is pregnant and has said she will terminate her pregnancy if the fetus has a life-threatening condition. The woman does not want to tell her sister about the test results, even though they mean that the sister's fetus may also be affected, because she believes that abortion is wrong. A physician or genetic counselor knows both women. Should the professional break patient confidentiality to tell the pregnant sister about the test results? "You [can] find yourself in a situation in which two individuals have competing moral interests," says Logan Karns, chair of the ethics subcommittee of the National Society of Genetic Counselors.[27] Because of these complex issues, organizations of genetic counselors recommend that individual counselors discuss the implications of testing carefully with people—and sometimes with their relatives as well—before testing takes place.

Chief among the legal issues raised by genetic tests are invasion of privacy (an important issue with regard to medical records in general) and discrimination, especially in insurance and employment. While many people see nothing wrong with insurers charging higher rates to those who choose unhealthy activities such as smoking, they feel it is unfair to penalize individuals for their genes, which they cannot choose or change. One invasion-of-privacy case related to genetic testing involved Lawrence Berkeley Laboratory (LBL), a California research facility funded by the Department of Energy. In February 1998, the U.S. Court of Appeals for the Ninth Circuit ruled that LBL had been wrong to test the blood of its employees for sickle-cell trait and other conditions without their informed consent. The court held that such action violated the protection against unreasonable search and seizure guaranteed by the Fourth Amendment. Because some tests were performed only on samples from blacks, Hispanics, or women, they also violated the 1964 Civil Rights Act. This was said to be the first case in which a federal appeals court recognized a constitutional right to genetic privacy.

National DNA databases such as the ones established in 1992 by the Department of Defense and in 1998 by the FBI have also been seen as threats to genetic privacy. In April 1996, John C. Mayfield III and Joseph Vlakovsky, both U.S. Marines with exemplary military records, were convicted in a court-martial of disobeying a direct order because they refused to give blood for storage in the Department of Defense database. The soldiers said that letting the government archive their genetic information violated their right to privacy, and they feared that the information might eventually be used to discriminate against them in some way. This was the first case to challenge the right of an employer to mandate sample donation for genetic testing. The case was later dismissed as moot because both men left military service before it was settled, but the military did agree to destroy blood samples at donors' request after the donors left the service. They still maintained their right to collect samples from active personnel, however.

Many people have told interviewers that they are afraid to be tested for defective genes they suspect they carry because they fear that insurers or employers will gain access to the results and then deny them or members of their families insurance and jobs. Some think that such discrimination has already occurred. For example, a 1996 poll of 322 members of families susceptible to inherited diseases found that 25 percent believed that they or other family members had been refused life insurance, 22 percent believed they had been refused health insurance, and 18 percent felt they had lost jobs because of their perceived health risks.

Of course, fears and beliefs are not the same as facts. In a survey conducted in 2000 by the American Management Association, only seven of 2,133 employers admitted to doing genetic tests on employees, when "genetic testing" was defined to include direct testing of genetic material and biochemical testing of blood samples but not taking of family histories. Similarly, Dean Rosen, senior vice president of policy for the Health Insurance Association of America, insisted in early 1999 that "the fears [of genetic discrimination] out there are just not reality."[28] But another insurance executive, who preferred to remain anonymous, advised people to "apply for insurance today, get [genetically] tested tomorrow" rather than the other way around.[29] Similarly, genetic counselors often recommend that people pay directly for genetic tests rather than charging them to their health insurance so that the insurance company will not find out the results of the tests. A 1996 study by Paul Billings, then at Stanford University, cited 455 cases of people being denied health care, insurance, jobs, schooling, or the right to adopt children because of a family history of inherited disease.

More than 40 states have enacted laws prohibiting insurers or employers from requiring applicants to take genetic tests, denying these groups access to genetic information, or preventing them from using such information in their decisions. Most other states are considering such laws. Laws of this kind, however, are often weakened by lack of a clear definition of genetic testing and, therefore, the kind of information that is covered. The laws also often do not cover genetic information that insurance companies obtain indirectly, for instance through family histories or through tests for gene products such as abnormal hemoglobin in the blood of people with sickle-cell trait. Finally, state insurance laws do not affect employers' self-funded plans, which in 1995 provided insurance for over a third of the nonelderly insured population.

The federal government has been less active than the states in attempting to ban genetic discrimination. Several pieces of antidiscrimination legislation have been introduced in Congress, but as of mid-2003, none had passed. President Bill Clinton signed Executive Order 13145, which prohibits genetic discrimination in federal employment, in 2000, but it does not apply to insurers or private employers.

Nonetheless, two existing federal laws do, or at least may, have an impact on genetic health testing and discrimination in employment. The Health Insurance Portability and Accountability Act (HIPAA), passed in 1996, is aimed mainly at preventing people from losing their health insurance when they change jobs, but it includes a provision that forbids health insurers issuing group plans to deny insurance on the basis of preexisting genetic conditions. It does not cover the 5 percent to 10 percent of Americans who have individual insurance plans, however, and it may not cover companies that insure their own workers, as opposed to those who buy group plans from insurance companies. It also does not necessarily prevent insurers from raising premiums on the basis of genetic information or protect the privacy of such information.

The Americans with Disabilities Act (ADA), passed in 1990, may also apply to healthy people who have inherited "bad" genes, if they can show that the results of genetic tests have made their employers perceive them as disabled. The federal Equal Employment Opportunity Commission (EEOC) filed its first suit against an employer for genetic discrimination in February 2001, alleging violation of the ADA. The suit was against the Burlington Northern Santa Fe Railroad, which had performed genetic tests without their knowledge on about 20 employees who had filed injury claims for carpal tunnel syndrome, damage to the hand and wrist usually caused by repetitive movement. The test was for a rare inherited condition that would predispose workers to the syndrome; if a worker had this condition, the railroad could claim that that employee's injury might not have been caused by work. One worker claimed that when he found out about the test and refused to provide a blood sample, the company threatened to fire him.

A little over two months after the EEOC filed its request for an injunction against the testing, the railroad agreed to settle the suit out of court. It admitted no wrongdoing (even the EEOC agreed that the company had not used the tests to keep anyone from working), but it promised to end all genetic testing, refrain from disciplining workers who had refused the test, and keep the existing test results confidential. Later it also agreed to pay the employees $2.2 million in damages.

Because of the railroad's settlement, the EEOC strategy was not tested in court. In *Bragdon v. Abbott*, a case brought before the Supreme Court in 1998, the court ruled that the plaintiff, who had tested positive for HIV but did not yet show any symptoms of AIDS, was protected by the ADA. A person carrying, say, BRCA1 or the gene for Huntington's disease would seem to be in a similar situation. In a June 1999 Supreme Court ruling on another ADA case (*Sutton v. United Air Lines*), however, Justice Sandra Day O'Connor wrote in her majority opinion, "We think the language [of the ADA] is properly read as requiring that a person be presently—not potentially or hypothetically—substantially limited [in a major life activity] in order to demonstrate a disability."[30] This would seem to dim hopes that the ADA will cover healthy people with genetic predispositions to illness.

Several decisions by lower courts also suggest that the ADA's protection against genetic testing and employer use of genetic information is limited at best. For one thing, commentators have pointed out, employees invoking the ADA would have to show, not only that an employer believed (correctly or otherwise) that they had a genetic defect, but that the employer regarded this as a disability (as defined by the ADA) and discriminated against the employees because of it. The ADA also does not prohibit the obtaining of genetic information, only certain uses of it. The law states that an employer may ask for any kind of medical information, genetic or otherwise, between the time of a job offer and the time a new employee begins work.

Discrimination based on genetic health testing has been an important issue in Britain, where the national health care system is increasingly supplemented by private insurance, as well as in the United States. The British government's Human Genetics Commission recommended in 1997 that there be a two-year moratorium on disclosure of the results of genetic testing and that a mechanism be established to evaluate the scientific evidence in support of particular genetic tests. In 2001 the insurance industry and the government agreed to a five-year extension of the moratorium, but a report from the Human Genetics Commission in the same year claimed that the industry's attempts at self-regulation were failing.

Efforts have also been made to protect against genetic discrimination in other European countries and in Europe as a whole. Norway has laws regulating genetic testing, for instance. Article 11 of the Convention for Human Rights and Biomedicine, adopted by the Council of Europe in November 1997, prohibits discrimination on grounds of genetic heritage, and the UNESCO General Conference's Universal Declaration on the Human Genome and Human Rights, also adopted in 1997, has a similar provision. As with laws opposing genetic discrimination in the United States, however, the meanings of terms such as *genetic testing* are not clearly defined in these rulings. The methods by which the declarations are to be enforced are also not clear.

Some commentators who are concerned about genetic discrimination in the United States say that new federal laws should be passed to protect genetic privacy and forbid employers and insurers from acquiring or using genetic information. Others say that, ultimately, genetic information is no different from any other medical information. Rather, they see the debate about insurers' use of genetic information as simply one aspect of the basic conflict between insurers (and employers who pay for employees' insurance) and insured regarding access to and use of all medical information.

As a number of writers have pointed out, this conflict is built into the nature of life and health insurance. Insurers can stay in business and make a profit only if they and the people they insure have equal knowledge of those individuals' risks of illness or death. If insured people alone know about a risk, the insurers say, they may take out large amounts of health or life insurance because they expect to need it. This "adverse selection" throws off the statistical methods by which insurance premiums are determined and drives up premiums for everyone. Harvie Raymond, director of managed care and insurance operations at the Health Insurance Association of America, insisted in 1995 that "he who assumes a risk should have the opportunity to evaluate that risk" and that laws barring insurance companies from obtaining genetic information "could create a great deal of havoc in the industry, causing costs to go up and fewer people to ultimately be able to afford coverage."[31] Conversely, if insurers know of a person's increased risk, they have a powerful financial incentive to deny insurance or charge very high premiums to that individual. This could make insurance unavailable to the very people who need it most.

Some commentators believe that, as genetic risks become easier to predict, this conflict will become soluble only by separating health care from the insurance system, probably by turning to a national health care system like the ones in Britain and Canada. Lori B. Andrews of the Chicago–Kent College of Law has predicted that "increasingly sophisticated genetic diagnostic tests may force a total rethinking of the concept of health insurance" and that the perceived injustice of genetic discrimination "will provide the impetus for the development of a national health system."[32] Another possibility, suggested by genetics expert Thomas Caskey, is to place people at high risk of genetic disease in a special insurance pool, as is sometimes done with otherwise uninsurable drivers. Alternatively, says Patrick Brockett, director of the Risk Management and Insurance Program at the University of Texas, insurers could be allowed to set rates on the basis of genetic tests, but people who do poorly on the tests could be given vouchers or other subsidies to help them pay for their insurance.

Another form of discrimination could stem from genetic determinism, the widely held belief that genetics—perhaps even single genes—are primarily or wholly responsible for such behaviors as violence, risk taking, and alcoholism. Although these behaviors may have a genetic association in certain cases, they are most likely determined by complex interactions between genes and environment. Nonetheless, if genetic determinism continues to be popular and genes associated with, say, violence are found, civil libertarians say, children with such genes might be put in special schools or otherwise segregated or treated differently from other children. Attributing violent behavior to genetics could also lead to a new variant of racism if "violence genes" prove to be more common in some racial or ethnic groups than in others. Perhaps the most likely and also most important danger of focusing on exclusively genetic causes of behavior such as violence, critics claim, is the diversion of attention and funding from correction of social causes of such behavior. "We know what causes violence in our society," says genetic discrimination expert Paul Billings. "[It is] poverty, discrimination, [and] the failure of our educational system."[33]

HUMAN GENE ALTERATION AND CLONING

On September 14, 1990, a four-year-old girl watching *Sesame Street* from her hospital bed made history. The child, Ashanthi deSilva, had been born with a mutant gene that made her body unable to produce an enzyme called adenosine deaminase (ADA). Lacking this protein, some of the cells in her immune system could not thrive. As a result, like other people with poorly functioning immune systems, she was easy prey for every microbe to which she was exposed. During all of her short life she had suffered from one infection after another.

Life changed for Ashanthi—and, potentially, for the world—on that September day. W. French Anderson, Michael Blaese, and their coworkers at the National Institutes of Health (NIH) had devised a technique for inserting a normal ADA gene into white cells (part of the immune system) taken from blood. When blood cells extracted from her earlier and treated with this method were reinjected into Ashanthi as she watched her hospital television, she became the first person to receive gene therapy, or injection of altered genes for the purpose of treating disease.

She was by no means the last. Although Ashanthi's therapy, which was repeated several times in the next two years, did not cure her disease, it was relatively successful. She had to continue taking an injected form of ADA, but the combined treatments allowed her to lead an almost normal life. Emboldened by this achievement, researchers went on to develop experimental gene treatments, not only for other inherited disorders such as hemophilia (which causes uncontrollable bleeding) and cystic fibrosis (which makes breathing difficult and increases susceptibility to lung infections) but for more common illnesses such as cancer, heart disease, and AIDS. They created a variety of techniques for altering the human genome, including correction of abnormal genes as well as insertion of normal ones. By early 2002, about 4,000 people had taken part in more than 450 gene therapy trials in the United States alone.

Unfortunately, with the possible exception of a handful of French children with another form of severe inherited immune deficiency, gene therapy has not completely cured any of the people who tried it. Furthermore, it has been blamed for at least one death and two cases of cancer—the latter in the same French children who had apparently been freed of their immune problem. Because of the safety issues raised by these alarming events, regulations governing gene therapy in the United States have been tightened considerably, and a number of experimental trials have been stopped either temporarily or permanently.

The person whose death was blamed on gene therapy was an 18-year-old Arizona man named Jesse Gelsinger, who had an inherited liver disease. He died on September 17, 1999, just three days after scientists at the University of Pennsylvania had injected a high dose of gene-carrying viruses into the blood vessel leading to his liver. His death shook the fledgling field of gene therapy and the agencies that regulated it: the FDA, the RAC, and NIH. In early 2000, after investigating the Gelsinger experiment and others, all three agencies sharply criticized gene therapy researchers for failing to use consent forms that fully warned prospective patients of possible dangers, failing to report bad reactions to treatments, and having close ties with commercial companies that stood to profit from the treatments if they succeeded (thereby creating a possible conflict of interest). In September 2000, Jesse Gelsinger's father also sued the researchers and institutions that had treated his son, but the suit was settled out of court after about two months.

Probably the greatest risk in gene therapy comes from the viruses that experimenters usually employ to carry new genes into cells. The scientists remove the genes that the viruses normally use to reproduce and cause disease, but the "crippled" viruses can still cause problems indirectly. In the case of the French children, whose illnesses were reported in September 2002 and January 2003, the viruses placed in cells from the children's bone marrow apparently unloaded their genetic cargo on top of a gene that controls cell growth, causing that gene to remain active longer than it should and overproduce certain blood cells. The result was a blood cancer much like leukemia. (The children are reportedly responding well to chemotherapy.)

Viruses caused a different difficulty for Jesse Gelsinger: His immune system apparently attacked them so vigorously that it killed him in the process. In some other experimental treatments, patients can safely be given a virus injection once, but researchers dare not repeat the process because severe reactions are likely to occur if the people's immune systems encounter the virus a second time. In still other cases, patients' immune systems have simply done what nature designed them to do—inactivated the viruses, including the added genes that the viruses carried. Experimenters are trying to develop safer and more effective ways to deliver genes, including ways that do not involve viruses.

Sometimes the genes themselves cause trouble. They may reach too few cells to be effective, may fail to integrate into cells' genomes, or may fail to make their proper proteins after they are in place. Alternatively, they may have unwanted effects along with desirable ones. For example, researchers in the late 1990s explored what seemed to be a very promising treatment for heart and blood vessel disease that used injections of a gene for a blood vessel growth factor to make new vessels grow to replace blocked ones. Cancerous tumors also benefit from the growth of blood vessels, however, and in at least one case the treatment apparently spurred the growth of a previously undetected cancer by providing the tumor with blood vessels.

Gene therapy researchers say that designing treatments is hard when so little is known about what most genes do and how they interact with each other and the environment. They counsel patience, even when experiments have occasional distressing outcomes. They point out that almost all medical treatments have had failures, even fatal ones, while they were being developed. Some critics question whether gene therapy will ever deliver the medical miracles its boosters have promised, but supporters such as W. French Anderson still believe that it will one day do so. "No other area of medicine holds as much promise for providing cures for the many devastating diseases that now ravage mankind," Anderson wrote in 2000.[34]

Meanwhile, gene therapy is by no means biotechnology's only contribution to medicine. Improved understanding of the ways that variations in human genes can affect the action and breakdown of drugs in the body is starting to help physicians tailor choices of drugs and dosages to individual patients to obtain maximum effectiveness with minimum side effects. DNA chips provide a new tool for diagnosing infections (by identifying key genes of bacteria and viruses) and other illnesses as well as for analyzing individual human genomes. Genetically engineered cells or bacteria grown in huge vats produce great quantities of proteins such as insulin that industry boosters say are cheaper and safer than their natural counterparts.

Biotechnology has also led to development of entirely new medicines, such as monoclonal antibodies, substances that home in on cancer cells while leaving normal cells alone. In a pattern that may be typical of the industry, monoclonal antibodies were first hailed as potential wonder drugs around 1980 but failed to live up to their promise and fell into disrepute for almost 20 years before finally achieving market success around 1999. Two such drugs, Rituxan and Herceptin, are now used regularly to treat certain types of cancer, and many others are being tested.

Most biotechnology-based drugs and treatments do not raise unusual ethical issues, but the potential social effects of human gene therapy may prove at least as important as its medical effects. Both supporters such as W. French Anderson and critics such as Jeremy Rifkin agree that Ashanthi deSilva's treatment represented not only a scientific advance but, as Anderson said at the time, "a cultural breakthrough, ... an event that changes the way we as a society think about ourselves."[35] They and many others, however, are still arguing about whether that change is for good or ill.

Few people question the morality of altering genes of human body cells to prevent or cure a life-threatening illness like Ashanthi's (although Germany, perhaps sensitive about the eugenics aspect of its Nazi past, forbids any alteration of human genes). After that, however, the ethical ground becomes shakier. Rifkin and other critics fear that, once the technical problems that presently limit human gene alteration have been solved, the definition of "disease" or "defect" will be stretched to include relatively minor problems (such as nearsightedness or obesity) or even mere differences (such as shyness or shorter-than-average height). If all these are engineered away, the result could be, at best, an undesirable loss of genetic diversity or, at worst, a new form of eugenics. Even W. French Anderson has said he feels strongly that gene therapy should be used only to treat serious disease.

Although governments might insist on, or at least strongly encourage, alteration of "defective" genes as a way to control health care costs, many of both critics and supporters of human gene alteration suspect that the demand for gene changes is more likely to arise from market forces than from government fiat. People might buy gene treatments for themselves in attempts to, say, grow larger muscles and make themselves into superathletes. Well-to-do parents might try to create "designer babies" by inserting genes likely to produce intelligence or physical beauty into their unborn children, just as they now purchase orthodontic treatments or special schooling.

Researchers warn that gene alterations could backfire, failing to produce desired effects or producing undesired effects or both. For example, an experimental gene alteration in mice has apparently made the mice more intelligent, as measured by increased speed in navigating mazes, but it has also left the animals in chronic pain. Even if alterations could be made predictable and safe, say critics such as University of Chicago bioethicist Leon R. Kass, the ability to specify genetic makeup could make parents think of their children more as goods ordered out of a catalog than as independent individuals.

Routine human gene alteration probably lies far in the future, but its potential for raising "slippery slope" ethical issues is well illustrated by a technique called preimplantation genetic diagnosis (PGD), which fertility clinics already employ. PGD does not alter genes, but it produces a similar effect by allowing prospective parents to select future children that either possess or lack genes for certain inherited traits. In this technique, a couple donate eggs and sperm to create fertilized eggs that will grow into embryos, just as they would for in vitro ("test tube") fertilization. When about a dozen embryos have developed to the size of a few cells, clinic workers remove a cell from each (at this stage, doing so does not harm the embryo) and examine the cell's genes. Based on the examination, they and the parents decide which embryos to implant in the mother's womb for further growth (several embryos are often implanted to be sure of obtaining a single full-term baby). PGD is most commonly used when both parents carry a gene that can cause a serious illness, such as sickle-cell anemia or early-onset Alzheimer's disease. Only embryos that do not contain the defective gene are implanted, guaranteeing that the couple's children will be free of the disease.

Few people argue with this use of PGD, but other applications are ethically murkier. In August 2000, for example, a couple in Denver, Colorado, gave birth to a son whom they had selected by PGD, not only to be free of Fanconi's anemia, an inherited disease that ran in both their families, but to be of a tissue type that was compatible with that of his older sister, who already had the disease. This compatibility meant that cells from his umbilical cord and bone marrow could be transplanted into his sister, potentially giving her a healthy blood system. (The transplant took place and was apparently a success.) A British couple obtained permission from their country's Fertilisation and Embryology Authority in 2002 to do something similar, but at about the same time the agency told another couple that they could not select an embryo to be a tissue match for a sick older sibling because the older child's disease was not inherited. This meant that, unlike the case with the other children, PGD would provide no medical benefit to the child on whom it was used.

A still more controversial use of (indirect) genetic selection also occurred in 2002, when a deaf lesbian couple deliberately chose a sperm donor with a strong strain of inherited deafness in his family in an attempt to ensure that they would produce a deaf child. They did not use PGD, but other deaf couples have said they would like to use that technique for the same purpose. Most hearing people find the idea of deliberately creating a child with what they see as a disability (an evaluation with which some deaf people strongly disagree) to be strange or even indefensible, but supporters such as Julian Savulescu of the Murdoch Childrens Research Institute in Britain say that if parents are allowed to select against inherited deafness when choosing an embryo to implant, they should have an equal right to select for deafness.

Present-day genetic selection has not been confined to medical conditions, just as many predict that future gene alteration will not be. Couples in Australia and the United States have used PGD for gender selection, for example. The American Medical Association (AMA) and the American Society for Reproductive Medicine have issued statements opposing selection of embryos for nonmedical reasons, but some parents apparently disagree.

The ethical questions raised by human gene alteration will become especially great if germ-line genes—those in the sex cells (the cells that become sperm and eggs), whose genetic information is passed on to offspring—are ever altered. Germ-line gene alteration has not yet been deliberately performed in humans, though it has been done in animals; all present gene therapy is intended to affect only somatic (body) cells.

Supporters of human germ-line gene modification say that it could eliminate inherited diseases completely, avoiding the necessity to choose or treat offspring in each generation. Critics, however, claim that PGD makes such risky action unnecessary. They also point out that even removal of a gene known to be associated with a serious disease might not be as clear-cut a good as it might seem, as the relationship between sickle-cell trait and resistance to malaria shows. By permanently deleting such a gene, therapists might unknowingly remove some characteristic that the human species will need at a future time. Altering genes associated with behavior would be even more likely to produce unpredictable effects. "Like Midas, bioengineered man will be cursed to acquire precisely what he wished for, only to discover—painfully and too late—that what he wished for is not exactly what he wanted," warns Leon Kass.[36]

Because of concerns like these, most people today, including most geneticists, feel that the human germ line should never be changed. In February 2000, for instance, nearly 250 concerned environmentalists and other leaders signed an open letter warning that human germ-line genetic engineering "is an unneeded technology that poses horrific risks" and urging that it be banned.[37] France, Germany, and India have outlawed human germ-line gene alteration, and the Council of Europe and the World Health Organization have said that they support a ban on it.

Nonetheless, germ-line alteration may already be occurring, at least inadvertently. In early 2001, a fertility clinic reported that, to avoid diseases carried in the mothers' mitochondrial DNA, some 30 children worldwide had been created by a technique in which an egg nucleus from the mother was transferred into an enucleated egg from a donor (with cytoplasm containing healthy mitochondria) and then fertilized with sperm from the father. All the cells of such babies appeared to contain DNA from all three people. This DNA is expected to be inheritable, so, in effect, the germ line was changed by addition of the donor's mitochondrial DNA, producing what the report called "the first case of human germ-line modification resulting in normal healthy children."[38] Some proposed applications of gene therapy to fetuses also present a small chance of altering the fetuses' germ cells, although that is not the treatments' purpose.

The idea of human cloning has generated even more heated debate than other proposed forms of human gene alteration, perhaps because it seems more likely than others to occur in the near future. It has haunted movies and novels as well as ethical discussions since genetic engineering began, and it took on new relevance after the appearance of Dolly, the cloned sheep, in 1997. Concern about human cloning was stirred several times during the early 2000s when individuals and groups, ranging from Italian fertility specialist Severino Antinori to a flying saucer cult called the Raelians, announced that they were about to produce or had produced a cloned child. As of mid-2003, no cloned babies have actually been proven to exist, but many people feel that, sooner or later, such a child will be born.

Objections to creating a cloned child—so-called reproductive cloning—begin with the practical one that, at least at present, such a procedure would be very risky to the prospective fetus. Dolly's was the only live birth in 277 tries, and only about 1 percent to 5 percent of subsequently created animal clones survived to birth. Many of these survivors proved to have defects, ranging from subtle to grotesque.

Ethical concerns about human cloning reach far beyond safety issues, however. Some critics say that cloning humans would be "playing God." Others fear that the parents of cloned children, or the children themselves, would regard the children as mere products rather than independent human beings. Producers of clones might expect to control the cloned children's lives completely or to see exact duplicates of themselves or whatever famous person they chose as the source of the clone. Some opponents of human cloning have pictured nightmare scenarios featuring, at one extreme, cadres of identical Hitlers (or, more benignly, Madonnas or Michael Jordans), or, at the other, armies of mindless slaves or even warehouses of headless bodies kept for possible organ donation.

Defenders of human cloning say that these fears are groundless. They point out that human cloning already occurs naturally in the form of identical twins, and they claim that an artificially produced human clone would simply be an age-delayed twin. They say that such a clone would be just as much a separate individual, with his or her own personality, as a twin is; indeed, the personality differences between a clone and his or her original would be greater than those between twins because they would be raised in different eras and therefore would be bound to have very different life experiences. Clones would be entitled to the same rights that other citizens have, including protection from slavery and forced organ donation. Cloning, these supporters say, could provide help to infertile couples who can reproduce in no other way.

Supporters, however, have been few. Most members of the public, legislators, and even scientists in North America and Europe today apparently feel that reproductive cloning should be made illegal, at least until safety questions are resolved, and perhaps permanently. The Council of Europe banned reproductive cloning on January 12, 1998, and at least two dozen individual nations, including Australia, Belgium, Britain, Denmark, Germany, India, Israel, Japan, the Netherlands, and Spain, have also banned or placed a moratorium on it. In the United States, California became the first state to outlaw reproductive cloning in January 1998, and six other states have followed. As of mid-2003, no federal law prohibits reproductive cloning—yet—but the FDA has claimed the right to regulate it and says that the agency will not give its approval to any reproductive cloning projects.

The U.S. House of Representatives has passed two bills banning human cloning, the most recent in February 2003, but companion bills have so far failed in the Senate. This has happened largely because, although legislators are virtually unanimous in wanting to ban reproductive cloning, they—and everyone else—are much more divided in feelings about a second type of human cloning, called research or therapeutic cloning. In this type of cloning, human embryos are created by somatic cell nuclear transfer, the same technique that produced Dolly, and allowed to multiply only up to the so-called blastocyst stage, a ball of about 200 cells. They are then destroyed so that certain cells within the ball, called embryonic stem cells, can be harvested. Embryos made by this kind of cloning would never be implanted into a uterus or allowed to develop into babies or even fetuses. Legislators and others disagree about whether laws should ban only reproductive cloning or all human cloning, including research cloning. The two bills that passed in the House banned all cloning.

Scientists have known about stem cells in the adult body for some 40 years. (These cells are also found in fetuses, newborns, and the blood of the placenta and umbilical cord.) Unlike most body cells, stem cells retain the ability to reproduce and do not completely specialize into particular types of tissue. They are the body's factories for making cells to replace those that wear out and die. Stem cells in the bone marrow generate both the oxygen-carrying (red) and the immune system (white) cells in the blood, for instance. Other stem cells replace skin cells and, to a limited extent, nerve cells. Adult stem cells have already been used in a few cancer therapies, and they are the reason that bone marrow transplants are an effective, though drastic, treatment for some diseases.

Stem cells in human embryos, however, were discovered only in 1998. James Thomson of the University of Wisconsin, Madison, and scientists at Geron, a California biotechnology company, purified the cells and developed a way to maintain them in the laboratory. They, like all other researchers on embryonic stem cells so far, obtained the cells from excess embryos created in fertility clinics and donated to research by the couples who had had them made. About a million such embryos are said to exist worldwide. Those not used in research are usually destroyed.

Most researchers in the field feel that embryonic stem cells have a far greater capacity to mature into multiple types of tissue than adult stem cells do. Boosters of embryonic cell research say that these cells could revolutionize medicine by providing sources of specialized cells to replace those damaged by disease and age, such as neurons to replace those destroyed by Parkinson's disease or Alzheimer's disease, or pancreatic islet cells to replace those destroyed by diabetes. Research cloning could allow larger numbers of such cells to be obtained than would be possible with leftover embryos. Furthermore, if replacement cells come from cloned embryos made from the body cells of a particular patient (as Dolly was made from a body cell of an adult ewe), they theoretically should not be rejected by the patient's immune system, making the taking of risky immunosuppressive drugs unnecessary. Research on cloned embryos and embryonic stem cells could also reveal much about how normal embryos develop and how diseases express themselves in individual cells, supporters say.

Other people, however, oppose the cloning of embryos for stem cell or other research. Some see all research on embryos as wrong because they believe that independent life and personhood begin at conception. Others, such as the American Association for the Advancement of Science, do not object to research on excess embryos from fertility clinics but dislike the idea of creating new embryos (cloned or otherwise) specifically for the purpose of experimenting on and destroying them. They feel that doing so comes too close to treating human life—even if it is not fully developed life—as a commodity, a mere "thing." Some feminists also fear that, if embryo cloning became widespread, a great demand would arise for donor eggs, leading to the exploitation of poor women, who, desperate for money, would consent to undergo the risky hormone treatments and surgical procedures necessary for egg donation.

Some critics also attack embryonic stem cell research on scientific grounds, saying that adult stem cells are likely to work just as well or that the promised use of cloned embryos to produce tissue transplants for individual patients probably would remain too expensive to be practical, even if it could be made to work from a technical standpoint. Some scientists also doubt whether use of stem cells from embryos cloned from a patient's cells would necessarily overcome transplant problems, because the embryo cells would contain different mitochondrial DNA, from the donor egg. Furthermore, studies of stem cells from cloned animal embryos have uncovered numerous genetic defects, suggesting that human embryonic stem cells might be dangerous to use as a medical treatment, perhaps inducing cancer, even if they could be produced.

Reaching a consensus on the ethics of research cloning and embryonic stem cell research has so far proved impossible on most occasions when it has been discussed. For example, the President's Council on Bioethics reported in July 2002 that, although all 17 members (which included scientists, ethicists, lawyers, and social scientists) favored a ban on reproductive cloning for both safety and ethical reasons, they could not agree on what to do about research cloning. Seven members wanted this type of cloning to proceed, but 10 others—a small majority—wanted it halted for four years to allow further public debate to take place. Some states that ban reproductive cloning permit therapeutic cloning, or at least embryonic stem cell research. Elsewhere in the world, Australia, Britain, Canada, China, and Scandinavia permit research cloning, but some EU member states do not. The same 1998 Council of Europe declaration that bans reproductive cloning also bans creation of human embryos for research purposes, but use of surplus in vitro fertilized (IVF) embryos is usually permitted.

Attempts to create a legal compromise regarding research cloning have been equally frustrating. On August 9, 2001, President George W. Bush announced that federal funding could be provided for research on human embryonic cell lines that already existed, but funding for creating or experimenting on new cell lines was forbidden. Bush claimed that more than 60 cell lines were available, but in mid-2003 the director of the NIH lowered that figure to about 12. Scientists who support embryonic stem cell research criticize the Bush measure because it restricts them to older cell lines instead of allowing them to use improved techniques to develop lines that might be better or safer. Those who believe that all research on embryos is wrong, on the other hand, feel that the measure does not go far enough. It does not affect research done by private companies, and it actually broadens the use of federal funding, which previously could not be applied to any research on human embryos.

Some scientists say that the whole issue of human cloning—for any purpose—may be moot because such cloning may never be practical or even possible. Researchers attempting to clone nonhuman primates such as monkeys have so far failed, apparently because certain proteins necessary for cell reproduction are damaged in the nuclear transfer process in primate eggs in a way that does not happen with the eggs of other mammals. A Worcester, Massachusetts, biotechnology company called Advanced Cell Technology attracted considerable media attention in November 2001 when it announced that it had cloned human embryos, but these embryos developed only up to the six-cell stage, and many scientists called the experiment a failure. By mid-2003, no cloned human embryos had been shown to develop even as far as the blastocyst stage, when stem cell harvesting could take place.

FUTURE TRENDS

No one knows what advances the coming "biotech century" will bring, but they are sure to be amazing, and their effects on society will be challenging, if not wrenching. There were good reasons why James Watson, the first head of the Human Genome Project, earmarked 3 percent of the project's budget for investigation of its ethical, legal, and social implications (ELSI). The implications of developments in biotechnology affecting plants and animals will be equally powerful.

One area likely to be of growing concern in the future is the relationship between biotechnology and bioterrorism. The letters containing anthrax spores mailed to U.S. media and politicians in October and November 2001 changed the view of bioterrorist attacks in the United States from a somewhat remote possibility to a reality, and from the beginning, both government and the public have feared that terrorists might use biotechnology to make deadly germs such as those that cause smallpox, bubonic plague, or ebola even more dangerous. The Soviet Union and pre-apartheid South Africa are known to have had bioweapons programs that included genetic engineering, and other countries may have done so as well. The "superbugs" created by these programs may still exist, waiting for clever or well-to-do terrorists to snap them up. Worse still, at least one FBI official has claimed that graduate or even undergraduate college microbiology students in a well-equipped laboratory could perform the same kinds of gene alteration that the programs carried out.

Two legitimate scientific experiments have shown how easy it would be for terrorists with laboratory experience to modify disease-causing microbes. In January 2001, researchers in Australia announced that about a year earlier, trying to make a contraceptive that would control rodent pests, they had inserted a gene for a protein found in the mouse (and human) immune system into viruses that cause mousepox, a weaker cousin of smallpox, and had found that the viruses became much more deadly than before. The altered viruses killed all the animals injected with them, even those that had been vaccinated against mousepox, by destroying their immune systems. "In the wrong hands, [other viruses altered in a similar way] could become biological warfare weapons of terrible proportions," said Bob Seamark, former head of the laboratory where the research was done.[39] Similarly, U.S. scientists announced in July 2002 that they had made a polio virus "from scratch," using gene sequences from a mail-order supplier and instructions downloaded from the Internet. They showed that their manufactured virus could cause disease in mice. Experts such as Gerald Epstein of the Department of Defense say that future scientists may need to work with governments to decide which experiments are too dangerous to publish or even, perhaps, to perform.

On the other hand, some legislators and officials, such as Connecticut Democratic senator Joe Lieberman, are encouraging government support of private biotechnology research in the hope that the industry can play a vital role in the country's defense against bioterrorism. Biotech companies have already invented detectors that accurately identify "signature" sequences of DNA belonging to several kinds of deadly bacteria in less than an hour, rather than in the days to weeks that traditional culturing methods require. Other firms are working on engineered vaccines that, unlike most existing ones, use only a small fragment of a protein found on the surface of a target virus or bacterium rather than whole weakened or killed microorganisms. A vaccine of this type for pneumonic plague, a form of plague that is highly contagious and spreads through the air, was tested on mice and found to be very effective, according to a report released in June 2003.

Despite the world's mixed feelings about its products, the agricultural biotechnology industry predicts continued growth. For example, a 2002 study by the Freedonia Group, an industrial market research firm, projects that U.S. sales of agricultural biotechnology products will advance 5.8 percent a year, reaching $2.8 billion by 2006. Transgenic crop acreage is predicted to increase 1.8 percent per year, to 102 million acres, during the same period. However, this rate of growth is considerably slower than that seen in the late 1990s. Continued bans of GM foods in Europe and elsewhere or increased disapproval in the United States, perhaps triggered by something like the StarLink incident in 2000 but more harmful, could cut into this rosy picture substantially. Environmentalists and others who oppose GM crops are sure to be watching for signs that these crops are causing the kinds of damage they have warned against.

The next generation of genetically engineered crops is likely to carry even greater benefits and risks than the present one, and the benefits are likely to be aimed at consumers rather than farmers. Most notably, scientists are experimentally engineering plants to produce vaccines and other substances useful in medicine. Such plant "pharming" is predicted to be a $12 billion industry by 2006 and $200 billion by 2013. If successful, it could considerably lower the prices of these medicines and also make them more easily available in developing countries. (Unlike conventional vaccines, for instance, plant-based vaccines would not require refrigeration, and they could be distributed as seeds and grown locally.) Because they would not require much acreage to show a profit, these new "pharm" crops would be easier for small farms to grow than present genetically engineered crops. Researchers are also engineering plants to produce fuel, biodegradable plastics, and other industrial products, potentially lessening dependence on oil.

On the other hand, biotechnology critics have expressed even greater concern about the possibility of pharmaceutical or industrial crops unintentionally entering the food supply than about present-day crops. A substance that cures illness in one person could cause it in another, they point out, and no one wants to find traces of, say, bioengineered plastic or gasoline in breakfast cereal. The possibility that such a disaster could happen was underlined in November 2002, when an ounce of corn engineered to produce a pig vaccine was detected in soybeans destined to become human food. The contamination did not reach the food supply in this case, but biotech opponents say that consumers might not be so lucky next time. In February 2003 the National Food Processors Association and several other food industry groups urged the government to place a moratorium on planting pharm crops until regulations could be shown to be strong enough to prevent any possibility of food contamination, and the USDA announced in March 2003 that it was strengthening inspection requirements for these crops.

Meanwhile, some scientists, particularly those working for national and international groups that are independent of large companies, are trying to tailor biotechnology to the needs of the developing world by engineering pest and disease resistance into crops that people there grow for food. For example, Kenyan scientist and biotechnology supporter Florence Wambugu has helped to engineer a yam to resist a virus that is a major pest of this crop, widely eaten in sub-Saharan Africa; the first trial crops of the new potato were harvested in late 2001. If more such crops are produced and donated to poor countries, claims that biotechnology brings little benefit to small farmers may be muted.

Legal and trade conflicts are likely to continue over patents and other protection of intellectual property in biotechnology, particularly in regard to biologically useful material discovered in developing countries and prepared for market by large companies in industrialized nations. Biotechnology companies, for their part, feel that they need and deserve patent protection in exchange for the considerable time, effort, and money they must spend in bringing new products to market, and they fear having their work exploited without compensation in countries where intellectual property laws are weak. They are sure to continue to demand the strengthening of such laws in international treaties and trade agreements. Nonwestern countries, on the other hand, will no doubt go on protesting these companies' patenting of materials and knowledge that originally came from those countries or of products that they expect people in those countries to buy.

Differing opinions about the morality of modifying and of patenting or "owning" types of living things, body parts, cells, and genes also will surely continue to clash. Some religious groups feel that claiming ownership or creation of living things usurps the role of God. Even among people who do not have strong religious beliefs, large numbers feel a deep moral unease about these subjects. The moral unease is deepest when the materials being patented or changed are human. This unease can only increase as the flood of information released by the Human Genome Project and related research grows.

As the complete sequences of more and more genomes are determined, the focus of biotechnology is shifting to the management and sale of genetic information. This focus has spawned new scientific disciplines called genomics and bioinformatics, which marry genetics and computer technology. These new sciences will increase the efficiency and scope of genetic engineering, for instance speeding the development of drugs and allowing the entire metabolisms of plants, animals, and even perhaps humans to be radically altered. They will also raise new ethical issues as some scientists or businesspeople begin to see (or to be accused of seeing) living things as mere bundles of information, as open to modification as computer programs.

DNA identification will surely continue to be a standard feature of certain types of court cases. Conflict about the extension of DNA databases beyond convicted felons is likely to grow, particularly if improvements in technology and funding remove the present backlog of untested samples. Another major terrorist attack in the United States could increase calls for a national database of all arrestees or even all citizens. On the other hand, continued conflicts about health insurance, the rise in identity theft, and concerns about government and corporate intrusions into privacy could spur demands that DNA samples from anyone except convicts be destroyed immediately after testing.

Unless major changes are made in the way health care is delivered and paid for in the United States, arguments about employers' and insurers' use, or misuse, of information from genetic tests are bound to grow as such tests become more widespread, more accurate, and more detailed. National laws forbidding genetic discrimination probably will be passed, but enforcing them may be difficult. This problem will become acute when precise and detailed genome readouts are routinely performed on babies at birth, which most likely will happen by midcentury at the latest. Such information will allow medicine to be individualized as never before and could lead to major improvements in both treatment and prevention of disease—if people are willing to abide by the health warnings they are given. On the other hand, it may also increase people's feeling that the entire path of their lives, at least in terms of health, was laid out unalterably in their genes before they were born. Both supporters and critics of biotechnology advances will have to work hard to avoid the pitfall of genetic determinism.

Biotechnology will continue to produce new drugs, vaccines, and other medical treatments. Some will surely save lives, perhaps large numbers of them. However, as Tom Abate, a California reporter who specializes in biotechnology, points out, new—and no doubt extremely expensive—drugs may not be health care's greatest need. Abate wrote in 2003, "We have an industry that has not yet turned a profit, trying to invent costly new medicines for a clientele that's having trouble paying for the things that are already in the medicine chest. This does not seem to be an encouraging trend."[40]

Biotechnology critics fear, furthermore, that drug companies' desire to make a profit may make them concentrate on products that enhance (or claim to enhance) the health and performance of basically well people rather than cure sick people (especially sick people in countries that cannot afford to pay high prices for medications). Such products might boost intelligence, endurance, or muscle strength, for example. They could prove to have unexpected side effects or be abused in various ways, such as providing an unfair advantage in athletic competitions.

The same may prove to be true with gene therapy and parents' genetic selection or alteration of their unborn children. Solving the technical problems besetting gene therapy may take several decades, but eventually alteration of genes in human body cells, probably increasingly done before birth, is likely to make inherited disease a thing of the past. It is also likely to be a part of treatments for more common conditions such as cancer and heart disease. "Health" and "disease" are parts of a continuum, however, and no one knows what degree of gene alteration society will accept. At least some wealthy parents, clandestinely if necessary, will probably attempt to enhance the genetic endowment of normal children. They may discover that producing another Albert Einstein or Marilyn Monroe is easier said than done, however, because environment shapes a child as much as genetics.

Genetic alteration or other biotechnological advances that greatly extend lifespans in developed countries or increase infant and youth survival in developing countries, desirable as these might be on an individual and humanitarian level, could exacerbate the world's already severe population problems. Furthermore, National Medal of Science winner Edward O. Wilson warns that a culture run by 150-year-olds is "likely to be a very conservative culture, one in which those who have survived and enjoyed longevity extension ... won't be revolutionaries. They won't be bold entrepreneurs or explorers who risk their lives."[41]

There is no technical reason why germ-line genes should be any more difficult to alter than the genes of body cells, but arguments about the ethics of making such alteration in humans are guaranteed to be more acute than those surrounding gene therapy or other alteration of body cell genes. Germ-line alteration, after all, goes beyond the individual to potentially affect the evolution of the entire species. Many people are sure to doubt the wisdom of taking on such an awesome responsibility.

What is banned today, however, may be permitted tomorrow. If alteration of human germ-line genes ever does become widespread, some commentators think it will ultimately change the very nature of the species. For instance, Princeton University geneticist Lee J. Silver predicts that, because only the wealthy are likely to be able to afford gene alteration for themselves or their children (at least at first), differences between those who can buy such treatments and those who cannot will eventually be so great that the two groups may become separate species. Supporters of germ-line alteration, such as DNA pioneer James Watson and researcher Gregory Stock of the University of California, Los Angeles (UCLA), say that such alteration could create "posthumans" with powers undreamed of today, such as supersenses, striking athletic ability, or perhaps even flight. "If we could make better human beings by knowing how to add genes, why shouldn't we do it?" Watson said in 2003.[42]

Reproductive human cloning will probably be banned in the United States and most other countries within the first decade of the 21st century. Nonetheless, unless the technical problems involved prove insurmountable, someone, somewhere, is almost sure to produce a cloned human child within a decade or so. Fears may die down when people realize that a cloned child is not much different from any other baby, as happened with in vitro fertilization once it became widespread. Some of the nightmare scenarios will fade when people realize that clones are not instant adults (as in some movies), automatons, or exact personality duplicates of their "parents." On the other hand, if the first cloned children prove to suffer from major deformities or health problems, further attempts at reproductive cloning may cease—voluntarily or otherwise.

Even if society eventually comes to accept reproductive cloning, such activity seems unlikely to become common. As a way to produce armies of either dictators or slaves, or even as a way to "resurrect" a lost loved one, most commentators agree, cloning simply will not work. It will probably be used only by a small number of infertile (including same-sex) couples and, perhaps, a few eccentrics who do not understand the interplay between genetics and environment. Laws defining family relationships will probably have to be modified if cloned human children come into existence, however, just as they have been altered in the past to accommodate surrogate motherhood, anonymous sperm donation, and other new reproductive technologies.

The fate of research cloning is harder to predict. It will probably depend on the overall progress of stem cell research. If further animal experiments and human tests fail to bear out the promise of stem cells, or if adult stem cells prove to be (or can be made) adequate for most proposed medical uses, the demand for using, or at least for creating, embryos for stem cell research will fall. On the other hand, if scientists develop a very effective new treatment for, say, cancer or Alzheimer's disease that requires embryonic stem cells, pressure from patients and their families may lead legislators to remove bans on research cloning and experiments on embryos. Laws on embryo research could also be affected by changes in laws about abortion.

In general, genetic engineering researchers will need to be more open with the media and the public if they want to regain the trust they have lost in recent years. Biotechnology companies, especially large ones, will also need to think harder about the actual and perceived impacts of their actions in order to avoid the kind of public disapproval and government bans that have blocked genetically modified foods in Europe. Both the industry and government regulators must show themselves very quick to detect and halt any environmental or health damage caused by biotechnology products. Biotechnology must also find more ways to benefit ordinary consumers, and particularly the world's disadvantaged people, if the industry wants to retain international respect and support.

In turn, the public—not to mention the public's media sources and political representatives—will need great improvements in education to deal with the ethical challenges of the biotech century. People must learn to grasp such things as the statistics of probability and risk or the complex interaction between genes and environment. Only education can help people sort through the hype and the nightmares presented by supporters and opponents of various techniques and reach reasoned conclusions about how humanity should use its wonderful and terrible new power to alter the essence of life. As Charles Weiner, emeritus professor of history of science and technology at the Massachusetts Institute of Technology, writes:

> *[Scientific] self-regulation is not adequate for today's urgent social and political choices about the directions, priorities, and limits to human and agricultural applications of genetic engineering and biotechnology. . . . They must be decided in the public arena and take account of concerns for social justice and moral values, as well as effects on health and environmental safety."*[43]

Leon Kass—along with many other thinkers—devoutly hopes that "human beings through their political institutions can exercise at least some control over where biotechnology is taking us."[44]

---

[1] "International Consortium Completes Human Genome Project," *Genomics & Genetics Weekly*, May 9, 2003, p. 32.

[2] Jeremy Rifkin, *The Biotech Century* (New York: Jeremy P. Tarcher/Putnam, 1998), p. xv.

[3] Unknown scientist, quoted in Edward Shorter, *The Health Century* (New York: Doubleday, 1987), p. 238.

[4] Steve Burrill, quoted in Tom Abate, "Celebrating 50 Years of DNA," *San Francisco Chronicle*, February 24, 2003, p. E4.

[5] "Climbing the Helical Staircase," *The Economist*, March 29, 2003, p. 3.

[6] Steve Burrill, quoted in Robert Winder, "Risky Business," *Chemistry and Industry*, January 20, 2003, p. 14.

[7] Michael Rogers, quoted in James D. Watson and John Tooze, *The DNA Story* (San Francisco: W. H. Freeman, 1981), p. 28.

[8] Norton Zinder, quoted in Burke Zimmerman, *Biofuture* (New York: Plenum Press, 1984), p. 141.

[9] Jeremy Rifkin, quoted in Paul Ciotti, "Saving Mankind from the Great Potato Menace," *California Magazine*, October 1984, p. 97.

[10] Levy Mwanawasa, quoted in Gavin du Venage, "African Nations Ban Biofood Aid Despite Famine," *San Francisco Chronicle*, August 23, 2002, p. A15.

[11] U.S.C. 35, Section 103(a).

[12] P. J. Federico, quoted in *Diamond v. Chakrabarty*, 447 U.S. 303.

[13] Council for Responsible Genetics, quoted in David B. Resnik, "DNA Patenting and Human Dignity," *Journal of Law, Medicine, and Ethics*, Summer 2001, p. 152.

[14] Jerry Caulder, quoted in Richard Stone, "Sweeping Patents Put Biotech Companies on Warpath," *Science*, May 5, 1995, p. 656.

[15] Anuradha Mittal, quoted in Peg Brickley, "Payday for U.S. Plant Scientists," *The Scientist*, January 21, 2002, p. 22.

[16] *John Moore v. Regents of California*, 51 Cal. 3d 120.

[17] European Court of Justice, quoted in "Europe Rules on Biotech Patents," *Chemistry and Industry*, November 5, 2001, p. S684.

[18] Unknown juror, quoted in Peter J. Neufeld and Neville Coleman, "When Science Takes the Witness Stand," *Scientific American*, May 1990, p. 46.

[19] Jamie Gorelick, quoted in "Fugitive Justice," *Nation*, March 3, 1997, p. 4.

[20] Rebecca Sasser Peterson, "DNA Databases: When Fear Goes Too Far," *American Criminal Law Review*, Summer 2000, p. 1219.

[21] Leroy Hood, quoted in Joel Davis, "Leroy Hood: Automated Genetic Profiles," *Omni*, November 1987, p. 118.

[22] Martha Volner, quoted in Geoffrey Cowley, "Flunk the Gene Test and Lose Your Insurance," *Newsweek*, December 23, 1996, p. 48.

[23] Nancy Wexler, quoted in Lauren Picker, "All in the Family," *American Health*, March 1994, p. 24.
[24] Francis Galton, quoted in William Cookson, *The Gene Hunters* (London: Aurum Press, 1994), p. 24.
[25] *Buck v. Bell*, 274 U.S. 200.
[26] *Skinner v. Oklahoma*, 316 U.S. 535.
[27] Logan Karns, quoted in "Genetic Testing Results: Who Has a 'Right' to Know?" *Medical Ethics Advisor*, May 2002, p. 52.
[28] Dean Rosen, quoted in Christopher Hallowell, "Playing the Odds," *Time*, January 11, 1999, p. 60.
[29] Anonymous insurance executive, quoted in Hallowell, p. 60.
[30] *Sutton v. United Air Lines*, 527 U.S. 471.
[31] Harvie Raymond, quoted in Seth Shulman, "Preventing Genetic Discrimination," *Technology Review*, July 1995, p. 17.
[32] Lori B. Andrews, quoted in Rick Weiss, "Predisposition and Prejudice," *Science News*, January 21, 1989, p. 41.
[33] Paul Billings, quoted in Bettyann H. Kevles and Daniel Kevles, "Scapegoat Biology," *Discover*, October 1997, p. 62.
[34] W. French Anderson, "The Best of Times, the Worst of Times," *Science*, April 28, 2000, p. 629.
[35] W. French Anderson, quoted in Joseph Levine and David Suzuki, *The Secret of Life* (Boston: WGBH Educational Foundation, 1993), p. 207.
[36] Leon Kass, "The Moral Meaning of Genetic Technology," *Commentary*, September 1999, p. 32.
[37] Bill McKibben et al., quoted in Richard Hayes, "The Quiet Campaign for Genetically Engineered Humans," *Earth Island Journal*, Spring 2001, p. 28.
[38] J. A. Barritt et al., quoted in Mark S. Frankel, "Inheritable Genetic Modification in a Brave New World," *Hastings Center Report*, March–April 2003, p. 31.
[39] Bob Seamark, quoted in Curtis Rist, "Genetic Tinkering Makes Bioterror Worse," *Discover*, January 2002, p. 76.
[40] Abate, "Celebrating 50 Years of DNA," p. E4.
[41] Edward O. Wilson, "A World of Immortal Men," *Esquire*, May 1999, p. 84.
[42] James Watson, quoted in "Design-a-Kid: Does Humanity Need an Upgrade?" *Christian Century*, May 17, 2003, p. 22.
[43] Charles Weiner, "Drawing the Line in Genetic Engineering," *Perspectives in Biology and Medicine*, Spring 2001, p. 208.
[44] Leon R. Kass, "A Reply," *Public Interest*, Winter 2003, p. 60.

CHAPTER 2

THE LAW AND BIOTECHNOLOGY

LAWS AND REGULATIONS

Hundreds of pieces of legislation, regulations, and policy statements relating to biotechnology and genetic engineering have been issued by the U.S. Congress, the congresses of the states, or other government bodies such as the National Institutes of Health (NIH) and the President's Council on Bioethics. This section details some of the best known and most important ones. They are grouped according to the five chief topics discussed in Chapter 1 (Agricultural Biotechnology and Safety, Patenting Life, DNA "Fingerprinting," Genetic Health Testing and Discrimination, and Human Gene Alteration and Cloning). Within each topic, they are arranged by date.

Agricultural Biotechnology and Safety

RECOMBINANT DNA ADVISORY COMMITTEE CHARTER

In accordance with Section 402 (b) (6) of the Public Health Service Act (42 U.S.C. 282), the NIH in October 1974 established the Recombinant DNA Advisory Committee (RAC), consisting of 15 members. At least eight of the committee members were to be experts in recombinant DNA research, molecular biology, or similar fields, and at least four were to be experts in applicable law, standards of professional conduct and practice, public attitudes, the environment, public health, occupational health, or related fields. According to an NIH description of the RAC posted on the Internet in 2000, a third of the committee's members "represent public interests and attitudes."[1] The committee's job is to advise the NIH director concerning the current state of knowledge and technology regarding recombinant DNA and to recommend guidelines to be followed by investigators in the field. The director is required to consult it before making major changes in existing NIH guidelines.

In the early years of its existence, the RAC had to approve most new types of genetic engineering experiments. Today, however, its approval is seldom required except in experiments involving humans (gene therapy); all clinical trials involving the transfer of recombinant DNA to humans that are funded by NIH must be registered and reviewed by the RAC. The RAC's charter was most recently renewed in 1997, and NIH wrote in 2000 that the committee still "serves a critical role in the oversight of Federally funded research involving recombinant DNA."[2]

COORDINATED FRAMEWORK FOR REGULATION OF BIOTECHNOLOGY

In June 1986, the federal Office of Science and Technology Policy published the Coordinated Framework for Regulation of Biotechnology, which divided regulation of genetic engineering research and technology among five agencies: the NIH, the National Science Foundation, the U.S. Department of Agriculture (USDA), the Environmental Protection Agency (EPA), and the Food and Drug Administration (FDA). The first two groups were to evaluate research supported by government grants. The latter three agencies were, and are, the chief regulators of the environmental testing and sale of biotechnology products, under the authority of several laws that were amended to include genetically altered organisms. The agencies consider whether such products are safe to grow, safe for the environment, and (if intended as human or animal food) safe to eat. State laws, such as seed certification laws, may also affect bioengineered products.

FEDERAL FOOD, DRUG, AND COSMETIC ACT (FFDCA)

First passed in 1938, the Federal Food, Drug, and Cosmetic Act (21 U.S.C. 9) has been amended to give both the EPA and the FDA control over certain biotechnology products. The FFDCA gives the EPA the right to set tolerance limits for substances used as pesticides on and in food and feed. This includes tolerances for residues of herbicides used on food crops genetically altered to be herbicide tolerant and for pesticides in food crops that produce such substances. The FFDCA gives the FDA the power to regulate foods and feed derived from new plant varieties, including those that are genetically engineered. It requires that genetically engineered foods meet the same safety standards required of all other foods.

The FFDCA was most recently amended in 1997. The FDA's current biotechnology policy under FFDCA treats substances intentionally added to food through genetic engineering as food additives if they are significantly different in structure, function, or amount from substances currently found in food. The agency has concluded, however, that many genetically altered food crops do not contain substances significantly different from those already in the diet and thus do not require FDA approval before marketing. This ruling has been criticized by those who believe that genetically modified food crops may threaten human health.

FEDERAL INSECTICIDE, FUNGICIDE, AND RODENTICIDE ACT (FIFRA)

The Federal Insecticide, Fungicide, and Rodenticide Act (7 U.S.C. 136) was passed in 1947 to regulate the distribution, sale, use, and testing of chemical and biological pesticides. After the EPA was established in 1970, it took over regulation of pesticides under FIFRA. FIFRA has been amended to include plants and microorganisms producing pesticidal substances, such as agricultural crops genetically modified to produce *Bacillus thuringensis* (Bt) toxin.

TOXIC SUBSTANCES CONTROL ACT (TSCA)

The Toxic Substances Control Act (15 U.S.C. 53), passed in 1976, gives the EPA the authority to, among other things, review new chemicals before they are introduced into commerce. Section 5 of TSCA was amended in 1986 to classify microorganisms intended for commercial use that contain or express new combinations of traits, including "intergeneric microorganisms," which contain combinations of genetic material from different genera, as "new chemicals" subject to EPA regulation under the act. The EPA believes that organisms containing genes from such widely separated groups are sufficiently likely to express new traits or new combinations of traits to justify being termed "new" and reviewed accordingly. The EPA handles this review under its Biotechnology Program. Altered microorganisms containing genetic material from two species in the same genus are not subject to regulation under TSCA.

EPA regulations under TSCA were amended on April 11, 1997, to tailor the general screening program for microbial products of biotechnology to meet the special requirements of microorganisms used commercially for such purposes as production of industrial enzymes and other specialty chemicals; creation of agricultural aids such as biofertilizers; and breakdown of chemical pollutants in the environment (bioremediation). According to the EPA, this change provides regulatory relief to those wishing to use these products of microbial biotechnology while still ensuring that the agency can identify and regulate risk associated with such products.

PLANT PEST ACT

The Animal and Plant Health Inspection Service (APHIS) is the agency within the USDA that is responsible for protecting U.S. agriculture from pests and diseases. Under the Plant Pest Act (7 U.S.C. 7B), originally passed in 1987, APHIS regulations provide procedures for obtaining a permit or for providing notification to the agency before introducing into the United States, either by import or by release from a laboratory, any "organisms and products altered or produced through genetic engineering which are plant pests or which there is reason to believe are plant pests." The law was amended in 1993, 1997, and 2000 to simplify requirements and procedures. The 1997 amendment made notification, rather than obtaining of a permit, sufficient for the release of most new types of genetically engineered plants into the environment.

Patenting Life

PATENT LAW IN THE CONSTITUTION

Article I, Section 8, of the U.S. Constitution gives Congress the power to enact laws relating to patents—that is, to "promote the progress of science and useful arts, by securing for limited times to authors and inventors the exclusive right to their respective writings and discoveries."

1952 PATENT LAW REVISION

Current patent law stems from Title 35 of the U.S. Code, which was revised on July 19, 1952. The law specifies the requirements for patentability and the procedure for obtaining patents. It also gives the Patent and Trademark Office the job of granting patents and administering patent regulations.

The parts of Title 35 of greatest concern to biotechnology are Sections 100 to 103, which describe the criteria that determine which inventions are patentable. Section 101 states that a patentable "process, machine, manufacture, or composition of matter, or any . . . improvement thereof" must be "new and useful." Section 102 further defines the requirement of novelty. Section 103 adds the qualification that a patent may not be obtained if "the subject matter [of the item to be patented] as a whole would have been obvious at the time the invention was made to a person having ordinary skill in the art to which said subject matter pertains." The third paragraph of Section 103, added later, specifically adds biotechnological processes, including gene alteration and production of cell lines, to the list of patentable items.

The 1952 patent law was cited in the landmark Supreme Court case *Diamond v. Chakrabarty*, which in 1980 allowed living things other than plant varieties to be patented for the first time. In his majority opinion on that case, Chief Justice Warren Burger referred to testimony accompanying the 1952 law in which Congressman P. J. Federico, a principal drafter of the legislation, stated that Congress intended it to "include anything under the sun that is made by man."

PLANT PATENT ACT

Traditionally, living things were held to be "products of nature" and thus not patentable. In 1930, however, Congress passed the Plant Patent Act (35 U.S.C. 15), which provides that "whoever invents or discovers and asexually reproduces any distinct and new variety of plant" may obtain a patent on it. The patent grants "the right to exclude others from asexually reproducing the plant or selling or using the plant so reproduced." Only asexual reproduction was mentioned in this act because, at the time, hybrids could not be made to "breed true" (reproduce sexually).

PLANT VARIETY PROTECTION ACT

By 1970, it had become possible to reproduce hybrid plant varieties sexually, that is, by seed. Congress therefore passed the Plant Variety Protection Act (7 U.S.C. 57), which extends the patent or patentlike protection of the Plant Protection Act to plants that could be reproduced in this way. The variety protected has to be new, distinct, uniform, and stable. Except for farmers, who are allowed to save and reuse seeds under certain circumstances, the act forbids unauthorized sexual reproduction of protected plant varieties. The act specifically excludes fungi and bacteria from its coverage.

DNA "Fingerprinting"

DNA IDENTIFICATION ACT

On September 13, 1994, Congress passed the DNA Identification Act (42 U.S.C. Sec. 14131). This act attempted to answer criticisms of the quality of forensic DNA testing by ordering the Federal Bureau of Investigation (FBI) to establish standards for quality assurance and proficiency testing of laboratories and analysts carrying out such testing. The FBI was supposed to set up a system of blind external proficiency testing (that is, testing done by an outside agency) for forensic DNA laboratories within two years, unless the agency concluded that such a system was not feasible.

The FBI did establish quality assurance standards and an accreditation program for forensic DNA testing laboratories, and it trains scientists to audit laboratories, but it does not directly oversee testing or require compliance except for those obtaining Department of Justice grants. (Individual DNA analysts who participate in the bureau's National DNA Index System [NDIS] are required to pass a proficiency test twice a year, but the bureau does not handle the testing.) Critics have questioned the wisdom of giving the FBI control over quality assurance for DNA testing because the bureau's own testing laboratory has been shown to have major quality control problems in at least two Justice Department investigations, one in 1997 and a second in early 2003.

The DNA Identification Act also authorized the FBI to establish a national database of DNA profiles from people convicted of crimes, samples recovered at crime scenes or from unidentified human remains, and relatives of missing persons who voluntarily donate samples. This database, the National DNA Index System (NDIS), opened in October 1998. NDIS is part of the Combined DNA Index System (CODIS), which also includes software that helps state and local laboratories coordinate their databases with the national one and each other so that law enforcement personnel can match their own felon and crime scene samples against others obtained anywhere in the country. By mid-2001, 137 laboratories in 47 states were participating in the CODIS program.

Genetic Health Testing and Discrimination

AMERICANS WITH DISABILITIES ACT (ADA)

Passed in 1990, the Americans with Disabilities Act (42 U.S.C. 12101–12111, 12161, 12181) was intended to increase disabled people's access to public spaces, communication, transportation, and jobs and to prevent discrimination against them in employment and other areas. Section 12102 of the act defines disability as meeting one of three criteria: (1) a physical or mental impairment that substantially limits one or more of the major life activities of an individual; (2) a record of such impairment; or (3) being regarded as having such impairment. People suffering from inherited diseases would surely be considered disabled, but it is not yet clear whether people who are presently healthy but are likely to develop an inherited illness later in life (as in late-onset diseases such as Huntington's disease) or have inherited a gene associated with increased risk of an illness such as cancer can be considered disabled. The Equal Employment Opportunities Commission ruled in 1995 that using genetic test results to deny employment to people in this category was discrimination under the ADA, but its ruling does not have the force of law.

HEALTH INSURANCE PORTABILITY AND ACCOUNTABILITY ACT

The Health Insurance Portability and Accountability Act, passed on August 21, 1996 (P.L. 104–191), was intended primarily to help people keep their health insurance when they change jobs. A paragraph in Section 701, "Increased Portability Through Limitation on Preexisting Condition Exclusions," forbids considering genetic information as a preexisting condition for insurance purposes unless a person is actually suffering from an inherited disease. Thus, for instance, a woman who is shown by a test to have a mutated form of the gene BRCA1, which is associated with an increased risk of breast and ovarian cancer, but who does not actually have cancer could not be denied insurance.

Human Gene Alteration and Cloning

THE PRESIDENT'S COUNCIL ON BIOETHICS REPORT

On February 27, 1997, Ian Wilmut and his colleagues at the Roslin Institute in Scotland startled the world by announcing that they had cloned a sheep, Dolly, from an udder cell taken from a six-year-old Finn Dorset ewe. Although intended primarily as an advance in agricultural biotechnology, this first cloning of a mammal from a mature body cell spawned fears that humans would soon be cloned as well. Indeed, several times in the early 2000s, individuals or groups announced that they were about to produce or had produced a cloned human child, although by mid-2003 none of these claims had been proved.

The possibility of cloning humans stirred up considerable concern among scientists, legislators, and the public. Most commentators agreed that, for safety reasons if nothing else, no attempts should be made to produce children by cloning, at least for the foreseeable future. Numerous bills to ban such attempts were introduced into state legislatures and Congress in the late 1990s and early 2000s, including two that passed in the House of Representatives in 2001 and 2003. Presidents Bill Clinton and George W. Bush also asked prestigious advisory bodies of scientists and ethicists to examine the issue and provide reports with recommendations. The National

Bioethics Advisory Commission produced such a report in 1997, the National Academy of Sciences in January 2002, and the President's Council on Bioethics in July 2002. None of these reports or recommendations has legal power, but they are of interest as scientific, ethical, and, to some extent, governmental policy statements.

The 17-person President's Council on Bioethics considered not only reproductive cloning—cloning for the purpose of creating a child—but a second kind of human cloning, called research or therapeutic cloning. In this type of cloning, cloned embryos would be allowed to develop only to a very early stage, when the embryo consists of about 200 cells. The chief purpose of research cloning is to allow scientists to harvest embryonic stem cells, which can develop into any type of tissue in the body and may prove useful in treatments for such illnesses as Parkinson's disease and Alzheimer's disease. Some people who oppose reproductive cloning accept research cloning, while others think that both types of cloning should be banned.

The President's Council on Bioethics was unanimous in urging the banning of reproductive cloning, but the group was almost evenly divided about research cloning. Ten members of the council, a slim majority, wanted a four-year moratorium placed on the process to allow more time for public debate. The remaining members felt that research cloning should proceed because of its great medical promise.

COURT CASES

A great deal of litigation has arisen over issues related to biotechnology and genetic engineering. In biotechnology, many cases have been patent disputes. In human genetics and genetic engineering, most have related to either DNA fingerprinting or genetic discrimination. This section describes some of the important court cases relating to patenting of living things (including products of the human body), forensic DNA testing, and genetics-based discrimination in insurance and employment (including eugenics laws).

Patenting Life

DIAMOND V. CHAKRABARTY, 447 U.S. 303 (1980)

Background

Ananda Chakrabarty, a scientist working for General Electric Corporation, modified a bacterium of the genus *Pseudomonas* so that it could digest crude petroleum, something no natural bacterium could do. He did so by getting the bacterium to take up four types of plasmids from other bacteria that could digest different components of crude oil, a technique that could be classified as genetic engineering but did not involve recombinant DNA (the individual plasmids were unaltered).

In June 1972, believing that his new bacterium would be useful in cleaning up oil spills, Chakrabarty applied for patents (in the name of General Electric) on the process of making the bacteria, a mixture in which they could be spread on water, and the bacteria themselves. The Patent and Trademark Office (PTO) granted the first two patent requests but rejected the third on the grounds that, as living things, the bacteria were "products of nature" rather than "manufactures" and thus were not patentable under U.S. law. The Patent Office Board of Appeals affirmed the PTO's decision. The U.S. Court of Customs and Patent Appeal (CCPA), however, reversed it.

The government appealed the decision to the U.S. Supreme Court in 1979. At first the court refused to hear the case, telling the CCPA to reconsider its decision in light of the comment that the court had made in a 1978 case, *Parker v. Flook*, in which the judges had said they believed they should "proceed cautiously when we are asked to extend patent rights into areas wholly unforeseen by Congress." The CCPA held its ground, however, and the Court accepted the case later in 1979. It was argued on March 17, 1980, under the full name *Sidney A. Diamond, Commissioner of Patents and Trademarks, v. Ananda M. Chakrabarty.*

Legal Issues

The question before the Supreme Court was technically a narrow one: interpreting the language of 35 U.S.C. 101 (part of the revision of patent law that Congress had passed in 1952) to determine "whether a live, human-made micro-organism is patentable subject matter"—that is, whether it was a "manufacture" or "composition of matter" as Congress intended the 1952 law to be construed. The root of the question, however, was deeply significant: Could a living thing—even one whose genes had been deliberately altered—be considered to be a human invention?

Traditionally, laws of nature, physical phenomena ("products of nature"), and abstract ideas have been held not to be patentable. In testimony accompanying the 1952 law, however, Congressman P. J. Federico, a principal drafter of the legislation, stated that Congress intended it to "include anything under the sun that is made by man."

Congress had already allowed the patenting, or at least a patentlike protection, of plant varieties in the Plant Patent Act of 1930 (which permitted plant breeders to keep exclusive rights to asexual reproduction of new varieties they developed) and the Plant Variety Protection Act of 1970 (which extended the protection to sexual reproduction of new plant varieties). House and Senate reports accompanying the 1930 act said that its purpose was to "remove the existing discrimination between plant developers and industrial inventors" because the acts of developing new plant varieties and new compositions of nonliving matter were conceptually equivalent.

The CCPA, in defending its decision, claimed that Chakrabarty's altered bacterium met the patent criteria of novelty, usefulness, and nonobviousness specified in the 1952 law (35 U.S.C. 101–103).

> *We look at the facts and see things that do not exist in nature and that are man-made, clearly fitting into the plain terms "manufacture" and "compositions of matter." We look at the statute and it appears to include them. We look at legislative history and we are confirmed in that belief. We consider what the patent statutes are intended to accomplish and the constitutional authorization, and it appears to us that protecting these inventions, in the form claimed, by patents will promote progress in very useful arts.*

The CCPA further maintained that the fact that the things that had been "manufactured"—that is, deliberately altered by humans—were alive made no difference for patenting purposes: "[There is] no *legally* significant difference between active chemicals which are classified as 'dead' or organisms used for their *chemical* reactions which take place because they are 'alive'" [emphasis in original]. Counsel for the PTO argued, on the other hand, that Congress did not want bacteria to be patentable because the Plant Variety Protection Act specifically excluded them. In general, the PTO said, Congress had not resolved the question of whether living things should be considered patentable.

Amicus curiae ("friend of the court") briefs filed by various groups brought up several other issues. Biotechnology critic Jeremy Rifkin's organization, the People's Business Commission (later the Foundation on Economic Trends), for instance, filed a brief urging that Chakrabarty's patent be refused because of the possible threats genetically altered organisms posed to human health and the environment. The brief claimed that granting the Chakrabarty patent would provide an incentive for commercial exploitation of the new gene-altering technology that was not in the public interest. The biotechnology company Genentech, conversely, filed a brief calling for acceptance of the patent, pointing to the 1978 relaxation of the NIH guidelines as evidence that most scientists no longer feared recombinant DNA technology.

Other briefs questioned the ethics of "owning" organisms or claiming to have made them, saying that only God could make living things. Representatives of the rapidly growing biotechnology industry, on the other hand, stressed that patent protection was vital to the industry's growth. They said that if patents on engineered organisms were not allowed, much of the knowledge being generated in the new field would remain hidden in the form of trade secrets rather than being revealed so that others could use it once the patents expired.

Decision

On June 16, 1980, the Supreme Court decided by a 5–4 vote that "a live, human-made micro-organism is patentable subject matter under [U.S.C.] 101. Respondent's micro-organism constitutes a 'manufacture' or 'composition of matter' within that statute." Chief Justice Warren Burger, writing the court's majority opinion, stated,

> *The patentee has produced a new bacterium with markedly different characteristics from any found in nature and one having the potential for significant utility. His discovery is not nature's handiwork, but his own; accordingly it is patentable subject matter.*

The court concluded that Congress had intended the patent laws to have wide scope. Burger rejected the patent office's argument that the fact that Congress had passed the two plant patent protection acts meant that it had not intended the original 1952 statute to cover living things. He cited congressional commentary on the 1930 act stating that the work of the plant breeder "in aid of nature" was patentable invention. The distinction, he said, was not between living and nonliving things but between unaltered products of nature and things that had been invented by humans. The exclusion of bacteria from the 1970 act, Burger said, meant only that bacteria were not considered to be plants, not that they were not considered to be patentable.

The patent office's second argument against Chakrabarty's patent was that microorganisms could not be patented until Congress expressly authorized such protection because Congress had not foreseen genetic engineering technology at the time it passed the 1952 law. Somewhat reversing the court's cautious position in *Parker v. Flook*, Burger wrote that he "perceive[d] no ambiguity" in Congress's language that would exclude genetically altered organisms.

Justice William Brennan wrote the minority opinion. He claimed that in the plant patent acts, Congress chose "carefully limited language granting protection to some kinds of discoveries, but specifically excluding others," including bacteria. "If newly developed living organisms not naturally occurring had been patentable under 101, the plants included in the scope of the 1930 and 1970 Acts could have been patented without new legislation," he stated. In conclusion, recalling the recommendation for caution that the court had expressed in *Flook*, Brennan commented,

> *I should think the necessity for caution is that much greater when we are asked to extend patent rights into areas Congress has foreseen and considered but not resolved. . . . [The majority's decision] extends the patent system to cover living material even though Congress plainly has legislated in the belief that [present law] does not encompass living organisms. It is the role of Congress, not this Court, to broaden or narrow the reach of the patent laws.*

Finally, the question of the dangers of genetically engineered organisms—the description of which Chief Justice Burger called "a gruesome parade of horribles"—was an issue beyond the court's competence, Burger concluded. "Arguments against patentability . . . , based on potential hazards that may be generated by genetic research, should be addressed to the Congress and the Executive, not to the Judiciary," he wrote. He also noted:

> *The grant or denial of patents on micro-organisms is not likely to put an end to genetic research or to its attendant risks. The large amount of research that has already occurred when no researcher had sure knowledge that patent protection would be available suggests that legislative or judicial fiat as to patentability will not deter the scientific mind from probing into the unknown any more than Canute could command the tides.*

Impact

Commentators such as patent lawyer Mitchel Zoler have claimed that from a strict legal standpoint, the *Chakrabarty* decision was "trivial law."[3] It broke no new legal ground, but rather provided only a minor clarification of existing patent laws. Furthermore, Donald Dunner, another patent lawyer, noted in the year following the decision that the ruling was "important but . . . not life or death of the [biotechnology] industry, and even had it gone the other way, it probably would not have been."[4] Even if altered microorganisms themselves had not been considered patentable, the processes of making them would have been, or their nature could have been kept hidden as trade secrets. Furthermore, the ease of altering bacteria suggested that patents, even once obtained, could be fairly easily circumvented by further alteration.

The psychological impact of this Supreme Court decision on both supporters and critics of biotechnology, however, was enormous. By the time the court made its ruling, dozens of patent applications for recombinant and other genetically engineered organisms, mostly bacteria, had been submitted to the PTO but had not been ruled upon. The PTO now felt free to begin granting these patents. The court's decision, which was widely publicized, gave a considerable boost to the biotechnology industry and encouraged those who were thinking of investing in it. Peter Farley of Cetus Corporation, a leading biotechnology business that had been among those filing amicus briefs in the case, said afterward that "the positive impact [of the decision] was in the Court's bringing genetic engineering as a commercial enterprise to the attention of the entire country."[5] The decision also ignited public discussion about whether patenting living things was ethical and to what degree a patent implied "ownership" of a life-form.

The PTO expanded the *Chakrabarty* ruling in 1987, extending patent protection to animals, cells and cell lines, body parts, plasmids, and genes, including human ones. After that, the only type of living thing that could not be patented was a whole human being.

JOHN MOORE V. REGENTS OF CALIFORNIA, 51 CAL. 3D 120 (1990)

Background

John Moore, a Seattle businessman, learned in 1976 that he suffered from a rare form of leukemia. He went to the Medical Center of the University of California, Los Angeles (UCLA), for treatment, where his case was assigned to David W. Golde, M.D. Informing Moore "that he had reason to fear for his life," Moore later alleged, Golde recommended removing Moore's enlarged spleen (an abdominal organ that makes blood cells) as a treatment for the disease. Moore consented, and his spleen was removed on October 20.

Moore alleged later that, unknown to him, Golde had noticed even before the operation that Moore's blood cells had an extraordinary ability to make immune system chemicals called lymphokines, which have potential commercial use as medical treatments because they stimulate production of immune cells that fight bacterial infections and cancer. Golde therefore "formed the intent and made arrangements to obtain portions of [Moore's] spleen following its removal." Furthermore, Golde called Moore back from Seattle for several additional treatments between 1976 and 1983, during which he took samples of Moore's blood, bone marrow, sperm, and skin.

During this same period, Golde worked with Shirley Quan, another UCLA researcher, to develop Moore's cells into an immortal cell line that was eventually named "Mo," after Moore. The two researchers and their employers, the Regents of the University of California, applied for a patent on their cell line in 1981, and the patent was granted in 1984. They then made lucrative contracts with Genetics Institute and the drug company Sandoz for use of the cell line.

At no time did Golde or anyone else involved in the research tell Moore about the cell line or the patent; indeed, Moore alleged, Golde repeatedly denied any commercial plans when Moore directly asked him about such a possibility. Moore nonetheless somehow found out about the extremely lucrative use to which his cells were being put (one estimate placed the potential value of products from the patented line at $3 billion). He filed suit in 1984, naming Golde, Quan, the Regents, and the drug companies as defendants. He claimed that he still "owned" his removed cells, at least in the sense that he had a right to say what use was made of them, and that he had a proprietary interest in any products that these or other researchers created with the cell line made from the cells.

A California superior court denied Moore's right to sue in 1986, but he appealed the decision, and an appellate court reversed it two years later, stating that "the essence of a property interest—the ultimate right of control — . . . exists with regard to one's human body." Golde, the university, and the other defendants appealed the case to the California Supreme Court, which heard it in July 1990.

Legal Issues

The California Supreme Court agreed to rule on whether John Moore had grounds to sue the defendants for using his cells in potentially lucrative research without his permission. The basic question was whether a person has an ownership right to body cells and tissues after they have been removed and, if so, whether this right entitles the person to compensation if others develop those cells or tissues into a commercial product.

"In effect," wrote Judge Panelli in the court's majority opinion, "what Moore is asking us to do is to impose a tort duty on scientists to investigate the consensual pedigree of each human cell sample used in research"—something that no court had ruled on before. This was an important issue. An Office of Technology Assessment report to Congress had noted in 1987,

> *Uncertainty about how courts will resolve disputes between specimen sources and specimen users could be detrimental to both academic researchers and the infant biotechnology industry, particularly when the rights are asserted long after the specimen was obtained. . . . The uncertainty could affect product developments as well as research. Since inventions containing human tissues and cells may be patented and licensed for commercial use, companies are unlikely to invest heavily in developing, manufacturing, or marketing a product when uncertainty about clear title exists. . . . Resolving the current*

> *uncertainty may be more important to the future of biotechnology than resolving it in any particular way.*[6]

The case also raised questions about what information a doctor has a "fiduciary responsibility" to give a patient in order to obtain truly informed consent before doing a medical procedure. To a lesser extent, it considered what financial or other responsibility secondary parties that have an interest in research done on a person's tissues but have no direct dealings with the person, such as a university or a drug company, bear to the person from whom the tissues came. Finally, as Judge Arabian wrote in a separate opinion concurring with the majority, the Moore case raised "the moral issue" of whether there is "a right to sell one's own body tissue for profit."

Decision

The California Supreme Court decided by a 5–2 vote that John Moore did not have a right of ownership over his tissues or cells once they had been removed from his body. "Moore's novel claim to own the biological materials at issue in this case is problematic, at best," Judge Panelli wrote in his majority opinion. Moore therefore could not sue under tort law for a conversion—essentially, a theft—of those parts. The court defined conversion as "a tort that protects against interference with possessory and ownership interests in personal property" and concluded that "the use of excised human cells in medical research does not amount to a conversion." Moore also had no direct right to income from the cell line patent because it represented invention on the part of the researchers, not any creative effort by himself, the court ruled.

In his majority opinion, Judge Panelli also expressed concern that allowing Moore to sue on the grounds of conversion

> *. . . would affect medical research of importance to all of society. . . . The extension of conversion law into this area [use of cells after they have legitimately been removed from the body] will hinder research by restricting access to the necessary raw materials. . . . Th[e] exchange of scientific materials [cell lines] . . . will surely be compromised if each cell sample becomes the potential subject matter of a lawsuit.*

Furthermore, Panelli wrote, "The theory of liability that Moore urges us to endorse threatens to destroy the economic incentive to conduct important medical research" because of the danger of lawsuits from disgruntled patients whose cells had been transformed into cell lines.

Although it rejected Moore's right to sue for conversion, the court found that David Golde had violated his "fiduciary duty" to Moore by not telling him about the proposed for-profit use of his cells. Golde's commercial plans represented a potential conflict of interest with his role as Moore's physician, and his failure to inform Moore of those plans denied Moore some of the facts he needed in order to give informed consent to the spleen operation. Judge Panelli wrote:

> *We hold that a physician who is seeking a patient's consent for a medical procedure must, in order to satisfy his fiduciary duty and to obtain the patient's informed consent, disclose personal interests unrelated to the patient's health, whether research or economic, that may affect his medical judgment.*

The court therefore ruled that Moore could sue Golde, at least, on the grounds that Golde had violated his fiduciary duty and that he had failed to obtain Moore's properly informed consent. Grounds for suing the other defendants were more dubious, though the court did not rule out the possibility of such suits.

All the judges agreed that Golde had violated his duties to Moore, but they had differing opinions about the larger question of Moore's ownership of his body tissues. In a separate concurring opinion, Judge Arabian wrote about "the moral issue" involved in the case. He claimed that to allow Moore to sue for conversion—in other words, to support Moore's claim that he owned his tissues after their removal and should have been paid for them—would be to "recognize and enforce a right to sell one's own body tissue for profit," an idea of which Arabian clearly did not approve. The judge wrote:

> *[Moore] entreats us to regard the human vessel—the single most venerated and protected subject in any civilized society—as equal with the basest commercial commodity. He urges us to commingle the sacred with the profane. He asks much.*

Arabian expressed fears that supporting Moore's claim would result in "a marketplace in human body parts" and stated that the state legislature should settle the question of whether such a situation was permissible.

Judges Broussard and Mosk expressed other viewpoints in separate dissenting opinions. Broussard supported Moore's right to sue for conversion because of the allegation that Golde had been planning his research before suggesting that Moore have his spleen removed. By not telling Moore about his plans, therefore, Golde had interfered with Moore's ownership rights to his cells *before* the cells had been removed. Broussard pointed out that the state's Uniform Anatomical Gift Act allowed people to specify donation of their organs for transplantation after their death and claimed that this fact supported the idea that people could say how they wanted donated parts of their body to be used. Broussard wrote:

The act clearly recognizes that it is the donor of the body part, rather than the hospital or physician who receives the part, who has the authority to designate, within the parameters of the statutorily authorized uses, the particular use to which the part may be put.

Unlike Panelli, Broussard did not feel that allowing occasional suits like Moore's (which had the unusual feature that the commercial usefulness of his cells had been discovered, and actively concealed from Moore, before the cells had been removed from his body) would put a damper on medical research with cell lines or that, even if it did, this was sufficient reason to deny Moore's right to sue. Because most of the value of the cell line patent lay in the researchers' work, Broussard suspected that the damages Moore would receive would be relatively small even if he won his suit.

Broussard's view of the effect the court's decision would have on the possible sale of cells or body parts was exactly the opposite of Arabian's. Broussard wrote:

Far from elevating these biological materials above the marketplace, the majority's holding simply bars plaintiff, the source of the cells, from obtaining the benefit of the cells' value, but permits defendants, who allegedly obtained the cells from plaintiff by improper means, to retain and exploit the full economic value of their ill-gotten gains free of their ordinary common law liability for conversion.

Judge Mosk also dissented from some of the majority's decisions and reasonings. If past judicial rulings did not cover ownership of body parts, Mosk saw no reason not to extend them:

If the cause of action for conversion is otherwise an appropriate remedy on these facts, we should not refrain from fashioning it simply because another court has not yet so held or because the Legislature has not yet addressed the question.

Mosk also felt that, although Moore had contributed no creative effort toward development of the patented cell line made from his spleen, he was nonetheless a kind of "joint inventor."

What . . . patients [like Moore] ... do, knowingly or unknowingly, is collaborate with the researchers by donating their body tissue. . . . By providing the

> *researchers with unique raw materials, without which the resulting product could not exist, the donors become necessary contributors to the product.*

Because of that contribution, Mosk said, Moore should be entitled to some compensation.

Mosk agreed with Broussard that a threat to medical research on cell lines was not a sufficient reason to deny Moore's claim, though he gave different reasons for his view. Secrecy and competition in the biotechnology industry had already severely inhibited the exchange of information and research materials, Mosk wrote. Furthermore, he claimed, researchers would know where their cells came from and whether proper consent for their use had been obtained if they engaged in "appropriate recordkeeping."

Above all, Mosk, like Broussard, felt that denying Moore's right to own his body parts would result in the human body being treated as a salable product and thus was morally reprehensible. Mosk quoted an earlier judicial decision that stated:

> *The dignity and sanctity with which we regard the human whole, body as well as mind and soul, are absent when we allow researchers to further their own interests without the patient's participation by using a patient's cells as the basis for a marketable product.*

Impact

John Moore eventually filed suit on the grounds that the court had left open. His suit was settled out of court.

The results of the Moore trial pleased biotechnology researchers, who had feared possible liability from working with cell lines or having to share revenue from them if Moore's right to sue was upheld. It disappointed those who disapproved of the patenting of living things or of tissues, cells, or genes taken from human beings. At the same time, it discouraged the establishment of a marketplace or "body shop" where human organs, tissues, or cells would be bought and sold; at least, they would not be sold by their original owners or while still residing in those owners' bodies.

Although the Moore case is often cited as establishing the principle that people do not own rights to their body tissues, the ruling does not apply outside California (or even necessarily to all cases within California). Some other states have different laws. In Oregon, for instance, a 1995 law (amended in 1997) specifically grants ownership rights over tissues and the genetic information derived from them to the people from whose bodies the tissues came. There have been no national rulings on this issue.

DNA "FINGERPRINTING"

FLORIDA V. ANDREWS, 533 SO. 2D 841 (FLORIDA FIFTH DISTRICT COURT OF APPEALS, 1988)

Background

In 1986, a number of women were raped, beaten, and cut in Orlando, Florida. The intruder entered their homes late at night, when they were alone, and evidence suggested that he had stalked them before the attacks to learn their habits. He covered each woman's head with a blanket or sheet, and only one of his victims, Nancy Hodge, saw his face. Police were able to obtain semen samples from Hodge and one other woman, a young mother.

Responding to a report of a prowler in early 1987, police captured Tommie Lee Andrews, a 24-year-old warehouse worker. Nancy Hodge picked out Andrews's picture from a photo lineup, and he was charged with her rape and that of the other woman who had provided a semen sample.

Tim Berry, Andrews's prosecutor, was reluctant to base his case entirely on Hodge's identification. A blood typing test suggested that Andrews could have committed the rapes—but so could 30 percent of the men in the United States. That summer, however, another attorney told Berry about the DNA "fingerprinting" technique that British geneticist Alec Jeffreys had invented a few years before. The technique had been used in numerous immigration and paternity cases and had just made its first appearance in a British criminal court. It allowed DNA from small samples of blood, semen, or other body fluids found at a crime scene to be compared with that from a suspect's blood at certain locations in the DNA molecule that differed considerably from person to person.

Berry sent the semen samples and a little of both Andrews's and Hodge's blood to Lifecodes in Valhalla, New York, one of the few laboratories in the United States then able to perform the test. The DNA in the semen samples matched that of Andrews but not Hodge. The Lifecodes analyst said the odds of the match occurring by chance (that is, of Andrews being innocent, yet still having DNA that matched that in the semen sample) were one in 10 billion—almost twice the population of the world.

Legal Issues

At the time of Andrews's trial, DNA profiling evidence had been used in only one other criminal case in the United States, and that case had not been a close parallel to that of Andrews. The British case, however, was similar. Jeffreys's test had shown a match between the DNA in semen found on two teenage girls who had been raped and murdered in Leicestershire and that in the blood of Colin Pitchfork, a 27-year-old baker. During the hunt for the killer, the police had taken the unusual step of asking all men between ages 13 and 30 in several villages—some 5,000 people—to voluntarily give samples of their blood for testing. Fearing detection, Pitchfork had persuaded a coworker to give blood in his stead, but the man bragged about it and was overheard. When questioned, he led the police to Pitchfork. Pitchfork confessed and was convicted of the crimes. DNA testing had also exonerated another suspect in the case, Rodney Buckland. Buckland had confessed to one of the killings, but his DNA did not match the semen sample, so he could not have been guilty.

The spectacular success of DNA "fingerprinting" in the Pitchfork case had been widely publicized. The test was still extremely new to forensics, however, and there was considerable question about whether it would meet the "*Frye* rule" (based on a 1923 case, *Frye v. United States*) by which many judges decided whether evidence from a new scientific technique would be admitted in a trial. In the *Frye* case, the court had ruled that a technique had to be "sufficiently established to have gained general acceptance in the particular field in which it belongs" before evidence based on it could be used.

Decision

Andrews's trial for the rape of Nancy Hodge took place in October 1987. In a pretrial hearing on October 19, Berry brought in an expert witness who testified that the technique on which Jeffreys's test was based, although new in the courtroom, was widely accepted in genetics and molecular biology laboratories. The judge agreed on this basis that DNA profiling met the Frye requirement, and he allowed the DNA evidence to be presented in Andrews's trial. When a Lifecodes expert brought up the one-in-10-billion statistic, however, the defense lawyers objected. Unprepared for the challenge, he withdrew the statistic. Without it, even the DNA evidence combined with Hodge's identification of Andrews apparently was not enough. The jury was unable to reach a verdict, and the judge declared a mistrial.

Andrews went on trial for the young mother's rape a few weeks later, however, and this time the prosecutors were able to provide legal backing for the use of the statistics that supported the DNA test results. In this case, furthermore, Andrews had left literal as well as genetic fingerprints behind. He was found guilty on November 6, becoming the first person in the United States to be convicted of a crime partly on the basis of DNA evidence. In addition, he was retried for Hodge's rape in February 1988, and this time, despite questions raised by the defense lawyers, he was convicted. His total sentences for the two convictions amounted to 100 years in prison.

Impact

Coming soon after the widely publicized success of DNA profiling in the Pitchfork case, the technique's usefulness in convicting Andrews caused prosecutors to turn to it eagerly in similar cases. It was hailed as "a prosecutor's dream," the greatest aid to identifying criminals since the development of fingerprinting a century before. Judges and juries began to accept it as well. Nonetheless, its validity usually had to be established in a separate Frye hearing for each case.

Alfaro v. Terhune, 98 Cal. App. 4th 492, 120 Cal. Rptr. 2d (2002)

Background

The state of California passed a law ordering the collection of DNA samples from felons convicted of certain crimes and establishing a database to hold the samples in 1988, almost immediately after forensic DNA identification began to be used in the United States. (By 2002, the law had been expanded to cover 13 violent crimes, and California's database was the largest in any state, containing more than 200,000 DNA profiles.) In 1997, a group of women death row inmates in the state, led (alphabetically) by Maria "Rosie" Alfaro, sued the California Department of Corrections to stop collection of their blood and saliva samples for the database, claiming that such collection was unconstitutional.

Legal Issues

The women's lawyers maintained that collection of samples without consent violated their protection against unreasonable search and seizure, guaranteed by the Fourth Amendment to the U.S. Constitution, as well as their right to privacy. The attorneys also argued that collection of samples from death row inmates served no useful purpose because the prisoners could not receive more severe sentences, even if DNA tests linked them to additional crimes.

Lawyers for the California Department of Corrections, on the other hand, pointed out that numerous previous court decisions had found collection of DNA samples from convicted felons to be constitutional. For example, in a 1995 case, *Rise v. Oregon*, the Ninth Circuit Court of Appeals held that sample collection was not cruel or unusual punishment. *People v. King*, a case heard in 2000 by the California Fourth District Court of Appeals, established that collection and use of DNA samples from convicted felons did not violate legal protection against double jeopardy because profiles are matched only against evidence from unsolved crimes. Another appeals court decision, *Shaffer v. Saffle*, established in 1998 that the taking of DNA samples did not violate Fifth Amendment protection from testifying against oneself because the Fifth Amendment applies only to verbal testimony.

Decision

A judge in Sacramento Superior Court granted an injunction against collecting the women's DNA in 1998, ruling that the law requiring the samples was constitutional but the sampling procedure was flawed. This halted collection of samples from all death row inmates for several years. When California's Third District Court of Appeals reviewed the case in May 2002, however, the appeals court judges unanimously upheld the constitutionality of the state's DNA databank law and its sampling procedures as well. The judges noted that all 50 states had similar laws and that other courts had upheld them. The prisoners' attorneys appealed to the state Supreme Court, but in August 2002 the higher court refused to consider the case, letting the appeals court ruling stand.

Impact

This decision provides further verification that collection of DNA samples from convicted felons is constitutional. However, it does not indicate whether collection of samples against their will from people arrested but not yet convicted, let alone people not charged with a crime, would be constitutional. Presumably such people would have a considerably greater expectation of a right to bodily privacy and, perhaps, freedom from searches and seizures than convicted felons would.

Collection of samples from California death row inmates resumed in September 2002. In that same month, Governor Gray Davis signed a new law authorizing the California Department of Corrections to use "reasonable force" (defined in the law as "the force that an objective, trained and competent correctional officer, faced with similar facts and circumstances, would consider necessary to gain compliance") in collecting blood and saliva samples from prison inmates or parolees convicted of qualifying offenses who refused to provide them voluntarily after being notified in writing that they must do so. Before passage of this law, a court order had been required before force could be used.

In November 2003, a panel of the Ninth U.S. Circuit Court of Appeals ruled unconstitutional a federal law requiring parolees to give blood samples for the CODIS database, calling the requirement an unreasonable search. It remains to be seen whether this ruling will affect prisoners.

HARVEY V. HORAN, 278 F. 3D 370 (2001, 2002)

Background

James Harvey was convicted of rape and forcible sodomy in Virginia in April 1990. His trial included DNA evidence that had been tested in 1989, soon after forensic DNA testing was introduced to the legal arena. Testing methods of the time were crude, and the results of the test were inconclusive.

Harvey has consistently claimed that he did not commit the crime. In 1993 he sought the assistance of the New York–based Innocence Project, which works to obtain new DNA testing for convicted felons who say that they are innocent and might be cleared by such testing. Lawyers for the project asked Robert F. Horan, Jr., the prosecutor in Fairfax County, for access to the rape kit used in Harvey's trial, claiming that new techniques of DNA testing might show that Harvey's semen did not match the DNA in the kit samples. Horan refused. Harvey's lawyers then sued Horan in a federal district court in Alexandria, Virginia, in July 2000, citing 42 U.S.C. 1983, a statute that permits citizens to sue state or local officials who allegedly violate their constitutional rights.

Legal Issues

From a technical standpoint, the chief issue was whether it was legally appropriate for Harvey (and, by extension, other felons) to use the Section 1983 statute to command access to DNA evidence. Since the statute refers to violation of constitutional rights, however, the more important underlying question was whether demands for access to and retesting of DNA evidence after conviction could be tied to any constitutional right. In addition to denying this possibility, Horan claimed that permitting convicts to demand DNA retesting would cost the state a crippling amount of money and time and would hamper the processing of new cases. Peter Neufeld, cofounder of the Innocence Project, denied this charge.

Decision

U.S. District Judge Albert V. Bryan, Jr., ruled on the case in Alexandria, Virginia, on April 16, 2001. He ordered Horan to provide the rape kit to Harvey's lawyers and, more important, stated clearly that, in his opinion, Horan had violated Harvey's civil rights—specifically, his right to due process under the Fourteenth and Fifteenth Amendments—and that convicted felons in general had a right to DNA testing that might prove their innocence. Bryan based his decision on a 1963 Supreme Court case, *Brady v. Maryland*, which held that prosecutors violated defendants' constitutional right to due process when they suppressed evidence. "Denying the plaintiff access to potentially powerful exculpatory evidence would result in . . . a miscarriage of justice," Bryan wrote. Contemporary news stories stated that Bryan's was the first such ruling by a federal judge in the country. Other judges in similar cases had not accepted the constitutional argument.

Horan appealed Bryan's decision, and the Fourth U.S. Circuit Court of Appeals reviewed it in January 2002. A three-judge panel voted to reverse the district court ruling. Writing the majority opinion, Chief Judge J. Harvie Wilkinson III stated that, in his opinion, Section 1983 was not the appropriate legal vehicle for Harvey to use in demanding the DNA evidence. He held that Harvey was trying to invalidate a final state conviction following a trial that, by Harvey's own admission, was fair in terms of the evidence available at the time.

Addressing the larger issue, Wilkinson wrote that he could see no grounds for claiming a constitutional right to access DNA evidence after a conviction. Like Horan, he feared that accepting Harvey's arguments would open the door to an endless round of convict lawsuits and expensive demands for rehearing as scientific techniques advanced. "Establishing a constitutional due process right to re-test evidence with each forward step in forensic science would leave perfectly valid judgments in a perpetually unsettled state," Wilkinson wrote. He also pointed out that other evidence, including testimony from a second man convicted of the rape, implicated Harvey.

Harvey appealed the circuit court panel's decision, asking that the case be reheard by all the circuit court judges (en banc). The court denied his request in March 2002. Judge Wilkinson again stressed that he felt that Section 1983 was inappropriate in this case. He stated that Harvey should have kept his appeals in the state court system rather than attempting to sue in federal court. Federal courts, he said, should not try to decide who, if anyone, has a right to demand DNA testing when Congress and numerous state legislatures were still wrestling with that issue. In a dissenting opinion, however, Judge Luttig maintained that, because avoidance of wrongful convictions was so important, convicts under at least some circumstances might well have a constitutional due process right to DNA testing.

Impact

Supporters of DNA testing for convicted felons hailed Bryan's decision as a landmark, even though it was not binding on other courts. The appeals court's rulings, however, dimmed their hopes for establishment of a constitutional right to DNA testing. At the same time, a number of states, including Virginia, were considering or passing laws providing some degree of testing access for convicts. In February 2002, in fact, just a month after the appeals court reversed the district court's decision, circuit judge David T. Stitt ordered Horan to turn over the evidence to Harvey's lawyers on the basis of a Virginia statute signed into law in May 2001, a month after Bryan's ruling. The new Virginia law gave people convicted of violent crimes, including death row inmates, the right to seek court orders for testing that might provide proof of their innocence. Ironically, when the new DNA test was finally performed, it confirmed Harvey's guilt.

Harvey was able to obtain his DNA evidence under the state law, his civil rights case presumably has become moot. However, the broader issue of a convict's constitutional right to demand DNA testing remains unsettled. Because this question has implications, not only for the use of forensic DNA databases but for application of the death penalty and possibly for criminal law in general, some case similar to Harvey's probably will be taken to the U.S. Supreme Court.

Genetic Health Testing and Discrimination

Buck v. Bell, 274 U.S. 200 (1927)

Background

In 1924, when she was 18 years old, Carrie Buck, a "feebleminded" (developmentally disabled) woman in the State Colony for Epileptics and Feeble Minded in Virginia, was ordered to be sexually sterilized under a newly passed state law that required such treatment for people living in state-supported institutions who were found to have hereditary forms of insanity or subnormal intelligence. Such sterilization was supposedly necessary to promote "the health of the patient and the welfare of society." Buck was deemed to be hereditarily feebleminded because her mother was of subnormal intelligence (she was confined in the same institution) and there were signs that Buck's illegitimate baby daughter was as well.

The Virginia law was typical of laws then existing, or soon to exist, in 34 states of the United States and several other countries, including Canada (some provinces), Britain, Germany, and the Scandinavian countries. These laws were based on the "scientific" doctrine of eugenics, which had been established in the late 19th century by British scientist Francis Galton and was widely accepted at the time of the Buck case. Galton and his followers believed that complex personality traits such as intelligence were inherited, and they claimed that the human race would be improved if groups such as the subnormally intelligent, habitual criminals, and the insane were prevented from reproducing—by force, if necessary. Such action, they said, would also save society considerable money by reducing the number of people who must be incarcerated and cared for at state expense.

Buck (or others acting on her behalf) sued the director of the institution to prevent her operation. The Virginia Supreme Court of Appeals supported the institution, but Buck's lawyers appealed, and the case came before the U.S. Supreme Court in 1927.

Legal Issues

Buck's suit alleged that she had been deprived of the right of due process of law guaranteed under the Fourteenth Amendment. She also claimed to have been denied equal protection of the laws because the Virginia law affected people inside institutions but not those outside. The underlying issue was whether the state had a right to forcibly prevent reproduction by people it deemed to suffer from inherited defects and therefore to be likely to produce undesirable offspring. "It seems to be contended that in no circumstances could such an order be justified," Supreme Court Justice Oliver Wendell Holmes noted in his majority opinion.

Decision

The Supreme Court ruled that the Virginia eugenics law did not violate either the due process clause or the equal protection clause of the Fourteenth Amendment, and it therefore denied Buck's right to sue. Justice Holmes wrote that Buck had been granted due process because the Virginia law contained "very careful provisions . . . [to] protect the patients from possible abuse," including requirement of a hearing attended by both the inmate and his or her guardian to determine whether the person was "the probable potential parent of socially inadequate offspring." "There can be no doubt that . . . the rights of the patient are most carefully considered" in this procedure, Holmes claimed, and all the steps of it had been followed with "scrupulous" care in Buck's case.

Holmes also wrote that Buck had not been denied equal protection, even though the law did not affect all citizens of subnormal intelligence equally. "It is the usual last resort of constitutional arguments to point out shortcomings of this sort," he complained. However, he said, "the law does all that is needed when it does all that it can, indicates a policy, applies it to all within the lines, and seeks to bring within the lines all similarly situated so far and so fast as its means allow."

Perhaps most important, Holmes defended the social as well as the legal validity of the eugenics law. He wrote:

> *We have seen more than once that the public welfare may call upon the best citizens for their lives. It would be strange if it could not call upon those who already sap the strength of the State for these lesser sacrifices, often not felt to*

be such by those concerned, in order to prevent our being swamped with incompetence. It is better for all the world if, instead of waiting to execute degenerate offspring for crime or to let them starve for their imbecility, society can prevent those who are manifestly unfit from continuing their kind. The principle that sustains compulsory vaccination is broad enough to cover cutting the Fallopian tubes. Three generations of imbeciles are enough.

Impact

The court's decision in *Buck v. Bell* not only upheld the constitutionality of at least some eugenics laws but demonstrated the prevalent thinking that found such laws both scientifically and morally justified. The *Buck* case was cited often in subsequent decisions about similar laws, such as *Skinner v. Oklahoma.* Ironically, however, Holmes and the others involved in the *Buck* case may have been wrong in the judgment that gave them jurisdiction over the family in the first place—the conclusion that the Bucks were "imbeciles." In a famous 1984 essay, scientist-writer Stephen Jay Gould reported that Carrie Buck had been reexamined in 1980 and was found to be of normal intelligence, and school records suggested that her daughter (who died in childhood) had been normal, too. The only "deficiencies" of the three generations of Bucks, Gould concluded, were that they were poor, uneducated, and violated contemporary sexual mores by giving birth to children out of wedlock.

SKINNER V. OKLAHOMA, 316 U.S. 535 (1942)

Background

Skinner, the plaintiff in this case, had not led an exemplary life. He was convicted of stealing chickens in 1926, followed by convictions for robbery with firearms in 1929 and 1934. In 1935, after his second armed robbery conviction, he was sentenced to the state penitentiary. Oklahoma law considered all three of Skinner's crimes to be "felonies involving moral turpitude" and stated that anyone convicted of two or more such felonies and sentenced to prison was a "habitual criminal." As such, the Habitual Criminal Sterilization Act, a 1935 Oklahoma eugenics law based on the belief that criminal tendencies were inherited, made him subject to sexual sterilization.

Skinner sued to prevent the operation. A jury trial confirmed that he could be sterilized without endangering his health. When the Oklahoma Supreme Court supported this decision, Skinner appealed on constitutional grounds. His case came before the U.S. Supreme Court in May 1942 and was decided on June 1.

Legal Issues

Skinner, like Carrie Buck before him, claimed that he had been denied equal protection under the Fourteenth Amendment. He also said that the Oklahoma law violated the Eighth Amendment because sterilization was cruel and unusual punishment. Underlying the particulars of the suit, as in the *Buck* case, was the question of whether eugenics laws as a whole were constitutional. The suit alleged that "the act cannot be sustained as an exercise of the police power, in view of the state of scientific authorities respecting inheritability of criminal traits."

Decision

The Supreme Court ruled that Skinner had been denied equal protection because the Oklahoma law exempted embezzlers from the sterilization penalty but included those (like Skinner) who were convicted of grand larceny. The distinction between the two crimes was a very fine one, having to do with exactly when the convicted person had formed the intent of stealing. The state as a rule was entitled to make such fine distinctions, Justice William O. Douglas wrote in his majority opinion, but when a penalty as severe and permanent as sterilization was involved, they became much more dubious. Douglas wrote:

> *Strict scrutiny of the classification which a state makes in a sterilization law is essential, lest unwittingly, or otherwise, invidious discriminations are made against groups or types of individuals in violation of the constitutional guaranty of just and equal laws. . . . When the law lays an unequal hand on those who have committed intrinsically the same quality of offense and sterilizes one and not the other, it has made as invidious a discrimination as if it had selected a particular race or nationality for oppressive treatment.*

Douglas noted that there was no reason for assuming that a tendency to commit larceny was inheritable but a tendency to embezzle was not:

> *Oklahoma makes no attempt to say that he who commits larceny by trespass or trick or fraud has biologically inheritable traits which he who commits embezzlement lacks. . . . We have not the slightest basis for inferring that that line [between larceny and embezzlement] has any significance in eugenics.*

In contrast to the decision in the *Buck* case, Chief Justice Harlan Fiske Stone concluded in a concurring opinion that Skinner had been denied due process because the Oklahoma law, unlike the Virginia one, did not provide for a hearing in which an individual could present evidence that he or she is not "the probable potential parent of socially undesirable offspring." Stone accepted that "science has found . . . that there are certain types of mental deficiency associated with delinquency which are inheritable" and affirmed the right of the state to "constitutionally interfere with the personal liberty of the individual to prevent the transmission by inheritance of his socially injurious tendencies." He insisted, however, that there was no proof that the traits of any entire legal category of criminals were inheritable, and individuals therefore had the right to a hearing to determine whether their particular "criminal tendencies are of an inheritable type." Skinner, he said, had been denied that right. "A law which condemns, without hearing, all the individuals of a class to so harsh a measure [as sterilization] . . . because some or even many merit condemnation, is lacking in the first principles of due process."

The most important difference between the *Buck* and *Skinner* cases lay in the court's comments about the underlying social issue of eugenics and the government's right to forcibly prevent certain people from reproducing. Justice Douglas wrote, "This case touches a sensitive and important area of human rights . . . a right which is basic to the perpetuation of a race—the right to have offspring." The case, he said, "raised grave and substantial constitutional questions."

> *We are dealing here with legislation which involves one of the basic civil rights of man. Marriage and procreation are fundamental to the very existence and survival of the race. The power to sterilize, if exercised, may have subtle, far reaching and devastating effects. In evil or reckless hands it can cause races or types which are inimical to the dominant group to wither and disappear. There is no redemption for the individual whom the law touches. Any experiment which the state conducts is to his irreparable injury. He is forever deprived of a basic liberty.*

Justice Robert H. Jackson used equally strong words in a second concurring opinion.

> *I . . . think the present plan to sterilize the individual in pursuit of a eugenic plan to eliminate from the race characteristics that are only vaguely identified and which in our present state of knowledge are uncertain as to transmissibility presents . . . constitutional questions of gravity. . . . There are limits to the extent to which a legislatively represented majority may conduct biological experiments at the expense of the dignity and personality*

and natural powers of a minority—even those who have been guilty of what the majority define as crimes.

Impact

Even though the court found for the plaintiff in this case, *Skinner* did not reverse the effects of *Buck*. It did not declare eugenics laws to be unconstitutional, scientifically invalid, or morally reprehensible. It did, however, express the sort of doubts about such laws that many people were beginning to feel. Its claim that reproduction was a basic right would often be cited in later cases.

Eugenics laws remained on the books of many states and countries until the 1970s. After the 1940s, however, they were seldom enforced. A combination of better understanding of heredity, which suggested that complex personality traits such as intelligence and criminality were determined as much by environment as by genetics, and a revulsion for eugenics principles triggered by revelation of Nazi genocide following World War II helped to make such laws unpopular.

NORMAN-BLOODSAW V. LAWRENCE BERKELEY LABORATORY, 135 F.3D 1260 (9TH CIR. 1998)

Background

While examining her medical records in the process of applying for workers' compensation in January 1995, an employee of Lawrence Berkeley Laboratory (LBL), a California research facility managed by the University of California and the U.S. Department of Energy, discovered that blood and urine samples she had given during a preemployment medical examination had been tested in several ways without her knowledge. The same proved to be true of other LBL employees.

After receiving a letter from the Equal Employment Opportunity Commission saying that they had grounds for a suit, Marya S. Norman-Bloodsaw and six other administrative and clerical employees of LBL filed suit against the laboratory and others in September 1995. The suit alleged that the laboratory had tested employees' blood and urine for syphilis, sickle-cell trait (in the case of black employees), and pregnancy (in the case of women). It was filed on behalf of all present and past Lawrence employees who had been subjected to the tests in question.

The U.S. District Court for the Northern District of California dismissed all the employees' claims in June 1997. They appealed, and the case went to the Ninth Circuit Court of Appeals in February 1998.

Legal Issues

As Judge Stephen Reinhardt stated in the appeals court's written decision,

> *This appeal involves the question whether a clerical or administrative worker who undergoes a general employee health examination may, without his knowledge, be tested for highly private and sensitive medical and genetic information such as syphilis, sickle cell trait, and pregnancy.*

The LBL employees claimed that the medical tests in question had been administered without their knowledge or consent and that they were not notified later of the tests or their results. They said that their federal and state constitutional right to privacy had been violated by the conducting of the tests, the maintaining of the test results, and the lack of safeguards against disclosure of the results to others because of the "intimate" nature of the conditions tested. They claimed violations of Title VII of the Civil Rights Act of 1964 because only African-American employees had been tested for sickle-cell trait and only women had been tested for pregnancy; furthermore, they alleged, later blood samples from black and Hispanic, but not other, employees had been tested again for syphilis. Finally, the employees claimed violations under the Americans with Disabilities Act because the tests were not related to their job performance or business necessity. They did not claim that LBL had taken any negative action regarding their jobs because of the tests or that it had revealed the test information to others, but they said that the laboratory had provided no safeguards against dissemination of that information.

In addition to asking for damages for themselves, the employees were suing, according to the court record,

> *. . . to enjoin [forbid] future illegal testing, . . . to require defendants . . . to notify all employees who may have been tested illegally; to destroy the results of such illegal testing upon employee request; to describe any use to which the information was put, and any disclosures of the information that were made; and to submit Lawrence's medical department to "independent oversight and monitoring."*

The defendants denied that any of the employees' claims had merit. The tests, they said, represented only a minimal intrusion beyond that which the employees had consented to as part of taking the medical examination and giving blood and urine samples. They claimed that signs posted in examination rooms, furthermore, had announced the tests and that employees had been asked about some of the items tested on a questionnaire that they completed as part of their examination. The questionnaire asked if the employees had ever had medical conditions including sickle-cell anemia, venereal disease, or (in the case of women) menstrual disorders. They therefore should not have been surprised at being tested for such conditions, the defendants claimed. The defendants also said that the testing had occurred so long ago that the statute of limitations for complaints about it had expired and that, in any case, the laboratory had stopped doing the syphilis tests in 1993 (because such tests turned out to be an economically inefficient way of screening a healthy population), pregnancy tests in 1994, and sickle-cell tests in 1995 (because most blacks were by then tested for sickle-cell trait at birth).

Decision

The circuit court of appeals reversed the district court's ruling that the statute of limitations prevented the plaintiffs from suing. The time limit began to run, the court said, from the time when the plaintiffs learned about the tests—1995—not the time when the tests were taken, as the district court had held. The circuit court said that the question of whether the plaintiffs knew or should have known that they were being tested would have to be settled at trial, but Judge Reinhardt maintained that the facts that the employees had consented to have a medical examination, give blood and urine samples, and answer written questions about certain medical conditions were "hardly sufficient" to establish an expectation of such testing. "There is a significant difference between answering [a questionnaire] on the basis of what you know about your health and consenting to let someone else investigate the most intimate aspects of your life," Reinhardt wrote. He also noted that "the record . . . contains considerable evidence that the manner in which the tests were performed was inconsistent with sound medical practice" in that the tests in question were not a routine or even an appropriate part of a standard occupational medical examination.

The appeals court upheld the district court's dismissal of the plaintiffs' claims under the Americans with Disabilities Act (ADA). First, Judge Reinhardt wrote, most of the testing at issue had occurred before January 26, 1992, the date on which the ADA began to apply to public entities. The employees tested after that date were tested as part of employee entrance examinations, which, unlike other examinations, are not required by the law to be limited to matters connected with a person's ability to perform job-related functions. The appeals court also disallowed claims under the ADA related to the way the employees' medical records were kept.

The appeals court supported the employees' right to sue on all the other grounds, however. It agreed that because of the tests' "highly sensitive" nature, they represented more than a minimal invasion of privacy beyond that involved in the medical examination that had been consented to. Judge Reinhardt wrote:

> *The constitutionally protected privacy interest in avoiding disclosure of personal matters clearly encompasses medical information and its confidentiality. . . . The most basic violation possible involves the performance of unauthorized tests—that is, the non-consensual retrieval of previously unrevealed medical information that may be unknown even to plaintiffs. These tests may also be viewed as searches in violation of Fourth Amendment rights. . . . The tests at issue . . . [also] implicate rights protected under . . . the Due Process Clause of the Fifth or Fourteenth Amendments. . . . One can think of few subject areas more personal and more likely to implicate privacy interests than that of one's health or genetic make-up.*

The court ruled that discrimination in violation of Title VII of the Civil Rights Act of 1964 and the Pregnancy Discrimination Act was shown by the fact that certain tests were given to some employees but not to others as a condition of employment or were given more often to some employees during employment. The unauthorized obtaining of sensitive medical information on the basis of race or sex in itself constituted an "adverse effect" as defined by the act, even though no negative effects on employment occurred. The plaintiffs therefore had grounds to sue on this basis as well, the court decreed.

The fact that the tests had been discontinued did not make the plaintiffs' claims moot, the appeals court ruled, because the laboratory could reinstitute the tests at any time. Plaintiffs suffered ongoing injuries from the tests in that the test records were still in the employees' files and could potentially affect employment decisions or be given to others, even though this had not so far happened.

Impact

The decision confirmed employees' right to medical privacy and to not have tests, including tests for inherited conditions such as sickle-cell trait, run on them without their informed consent. In describing the decision, *U.S. News & World Report* writer Dana Hawkins said it represented "the first time a federal appeals court has recognized a constitutional right to genetic privacy."[7] The fact that Judge Reinhardt specifically mentioned genetic make-up in connection with privacy rights may prove particularly important to those concerned about privacy and discrimination related to genetic testing.

Partly because of this suit, Department of Energy contractors are now required to give employees a "clearly communicated" list of all medical examinations they will be expected to take, the purpose of the exams, and the results of the tests.

Bragdon v. Abbott, 97 U.S. 156 (1998)

Background

In September 1994, Sidney Abbott visited her dentist, Randon Bragdon, in Bangor, Maine. In filling out a patient registration form, she indicated that she had been infected with HIV since 1986, although she had not yet developed any symptoms of AIDS. Bragdon examined Abbott, found that she had a cavity, and informed her that his policy was not to fill cavities of HIV-positive patients in his office. He then offered to do the work at a nearby hospital (where he felt he could take better precautions to protect himself from infection) if Abbott was willing to pay the extra cost of using the hospital's facilities. She declined.

Abbott sued Bragdon for violating her rights under Title III of the Americans with Disabilities Act (ADA) by not treating her, since places of "public accommodation" defined in that section include the "professional office of a health care provider." Bragdon's lawyer pointed out that the act stated that people could refuse to do something for an individual that was otherwise required by the act if they could show that "said individual poses a direct threat to the health and safety of others," and he claimed that Abbott fit that description. Abbott's attorney disagreed, pointing out that the Centers for Disease Control and Prevention (CDC) and others had written guidelines describing procedures by which dentists could treat people with HIV infection.

A district court ruled that Abbott's HIV infection satisfied the requirements of disability under the ADA and that Bragdon had not proved that treating her would put his health at risk. The Court of Appeals affirmed both of the district court's rulings. Bragdon then appealed to the U.S. Supreme Court, and the case came before the Court in June 1998.

Legal Issues

The Court agreed to rule on the following points: (1) whether Abbott, as an asymptomatic person with HIV infection, was disabled as defined by the ADA, and (2) whether sufficient evidence had been provided to show that Bragdon's health would have been endangered by treating her.

The underlying issue of the case for those concerned about genetic discrimination was whether a person who was presently healthy but likely to become ill later could be considered disabled under the ADA, since this description fitted healthy people whom tests revealed to have a genetic susceptibility to a disease. A factor likely to affect this issue was the section, or "prong," of the ADA under which Abbott claimed disability. The act defines disability as having to meet one of three criteria:

1. a physical or mental impairment that substantially limits one or more of the major life activities of an individual;
2. a record of such impairment; or
3. being regarded as having such impairment.

Abbott claimed disability under the first prong, saying that HIV infection limited her in the major life activity of reproduction and childbearing. By contrast, the Equal Employment Opportunity Commission (EEOC) had stated as policy in 1995 that healthy people with genetic predispositions were covered under the act's third prong.

Decision

In a 5–4 decision, the Supreme Court ruled that Abbott's HIV infection rendered her disabled according to the ADA's first criterion, that of limitation of a major life activity. Justice Anthony Kennedy in his majority opinion stated that reproduction was, without question, a major life activity and that Abbott was substantially limited in her pursuit of it, since the unprotected sex necessary to conceive a child would put her partner at significant risk of infection. Furthermore, if she did become pregnant, the child would also have a good chance of being infected. Kennedy also noted that, even though HIV infection did not produce obvious symptoms for years, it caused steady damage to the blood and immune systems and thus was "an impairment from the moment of infection."

In writing of the possible threat to Bragdon's health, Justice Kennedy noted that the ADA defined a direct threat to be "a significant risk to the health or safety of others that cannot be eliminated by a modification of policies, practices, or procedures or by the provision of auxiliary aids or services." The basic question, Kennedy wrote, was whether Bragdon's actions were reasonable in light of the medical evidence available to him at the time. Kennedy pointed out that, on the one hand, Bragdon had not produced medical evidence to show that he would have been any safer treating Abbott in a hospital than in his office. On the other hand, the CDC and other similar guidelines do not necessarily say that dentists will be safe while treating HIV-infected patients if they follow the recommended procedures. Some such guidelines do say that risk is minimal if proper precautions are followed, and medical testimony had been offered to this effect as well in the previous trials, but Kennedy noted that Bragdon may not have had this information at the time he treated Abbott. Kennedy ordered the Court of Appeals to reconsider the evidence supporting Bragdon's estimation of his health risk.

Impact

According to the summary of a February 1999 workshop held jointly by the National Human Genome Research Institute and the Hereditary Susceptibility Group of the National Action Plan for Breast Cancer to discuss the implications of the *Bragdon v. Abbott* decision for healthy people diagnosed with a genetic susceptibility to breast cancer (and, presumably, other gene-related illnesses), the Court's ruling "both excited and unnerved" those who hoped that the ADA could be interpreted in a way that would cover such people.[8] Chief Justice William Rehnquist, in fact, addressed this possibility when he wrote in his dissenting opinion that Abbott's argument, "taken to its logical extreme, would render every individual with a genetic marker for some debilitating disease 'disabled' here and now because of some possible future effects." On one hand, Rehnquist's words indicate that such reasoning is possible; on the other, he obviously was expressing disapproval of the idea.

In the workshop, law expert Paul Miller pointed out several hopeful signs in the *Bragdon* decision. The fact that the Court considered HIV infection an actual disability from the beginning because of its effects on immune cells, even though no obvious symptoms of illness appeared, suggested that genetic tendencies, which may also cause physical or chemical changes in cells without producing visible symptoms, might be similarly classified. Furthermore, since reproduction had been affirmed as being a major life activity covered by the ADA, people with inherited defects could argue, as Abbott did, that their ability to reproduce was limited because they risked passing their condition on to their offspring and thus endangering those offspring's health. Third, Miller said, the court had relied heavily on an EEOC policy ruling in determining that asymptomatic HIV infection qualified as a disability under the ADA. This added weight to other EEOC rulings, including the one about genetic predisposition.

On the other hand, it was not clear how broad the effect of the Court's ruling would prove to be. The ruling did not explicitly consider any life activity other than reproduction, for instance, so it might not cover, say, a post-menopausal woman or a gay man with asymptomatic HIV infection. More important, since Abbott claimed disability under the first prong of the ADA's definition, the Court decision did not illuminate the question of who would be included under the third prong, which many commentators feel is the one most likely to cover healthy people with genetic predispositions. Miller noted that many legislators are not supportive of the third prong, and a later speaker, Sharon Masling, said that the courts also have been interpreting it narrowly.

The Law and Biotechnology

SUTTON V. UNITED AIR LINES, INC., 527 U.S. 471 (1999)

Background

In 1992, Karen Sutton and Kimberly Hinton, twin sisters, applied to United Air Lines for employment as global commercial airline pilots. They met all of the airline's criteria for this job except one: the minimum vision requirement, which called for uncorrected visual acuity of 20/100 or better. Both sisters were severely myopic (nearsighted), with uncorrected vision of 20/200 in their better eyes and 20/400 in their poorer ones. This degree of impairment, if uncorrected, would keep them from driving, shopping, and carrying out other everyday activities. With corrective lenses, however, both had vision of 20/20 or better. The airline refused to hire the women because they did not meet the minimum requirement for uncorrected vision. They sued United, claiming that the airline had discriminated against them in a way that violated the Americans with Disabilities Act (ADA).

Legal Issues

The sisters stated that they had a disability that "substantially limits a major life activity," as required by the ADA, because of the limits that their uncorrected vision would place on working and other activities. They cited guidelines issued by the Equal Employment Opportunity Commission (EEOC) in support of their claim that their disability should be evaluated as a disability under ADA in its uncorrected rather than its corrected state. They also argued that United "regarded" them as having such a disability, whether they actually did or not, which would entitle them to protection under the third prong of the ADA.

United claimed that the women did not have a disability because glasses could easily restore their vision to normal levels. It also stated that the company did not regard them as being substantially limited in the activity of working because it did not exclude them from a large class of jobs, only from the specific job of global airline pilot.

This case, like *Bragdon v. Abbott*, had indirect implications for the question of whether currently healthy people diagnosed with a genetic susceptibility to disease are protected by the ADA. If the courts accepted the argument that the sisters' disability should be measured in terms of their uncorrected vision, this would suggest that healthy people with genetic susceptibilities might be protected by the ADA because they were likely to be disabled in the future, even though they were not so in the present, just as the sisters might be disabled without glasses even though their vision was normal when they wore corrective lenses. The question of whether being excluded from a specific job was sufficient to demonstrate that an employer regarded someone as disabled could also relate to employer decisions about whether to hire or retain people with flaws in their genetic makeup.

Decision

A district court heard the sisters' suit in 1996 and dismissed it, saying that they had failed to state a claim upon which relief could be granted. The court held that, because the women's vision could be corrected to normal, they were not actually disabled. It also ruled that, because United had denied them access only to one specific type of job, it had not demonstrated that it regarded them as substantially limited overall in the life activity of working. The 10th Circuit Court of Appeals affirmed the district court's judgment in 1997. The sisters' lawyers then appealed to the U.S. Supreme Court, which agreed to hear the case because the lower courts' decisions conflicted with other decisions holding that disabilities should be evaluated without consideration of whether they could be or were corrected.

The Supreme Court issued its ruling on June 22, 1999, with Justice Sandra Day O'Connor writing the majority opinion. The high court affirmed the ruling of the lower courts. O'Connor concluded, "we think the language [of the ADA] is properly read as requiring that a person be presently—not potentially or hypothetically—substantially limited [in a major life activity] in order to demonstrate a disability." By this standard, she said, the sisters were not disabled. She cited, for instance, a statistic on the number of disabled Americans included in the text of the ADA, saying that the figure would have been much larger if Congress had intended to include people with correctable as well as uncorrectable disabilities.

O'Connor also held that the sisters failed to qualify under the "regarded as" prong of the ADA because the position of global airline pilot did not represent a class of jobs. "The inability to perform a single, particular job does not constitute a substantial limitation in the major life activity of working," she wrote, and the fact that the airline refused to consider the sisters for this specific job therefore was not sufficient evidence that it regarded their (uncorrected) poor eyesight as substantially limiting their general ability to work. The women might, for example, have qualified as copilots, regional pilots, or pilot trainers. O'Connor pointed out that "the ADA allows employers to prefer some physical attributes over others and to . . . decide that some limiting, but not *substantially* limiting, impairments make individuals less than ideally suited for a job."

Justices John Paul Stevens and Stephen Breyer dissented from the majority opinion. They held that Congress had intended the ADA's coverage to be broad and, therefore, that the possibility of correction should be disregarded when determining whether a person is disabled as defined by the law. To Stevens, "it [was] quite clear that the threshold question whether an individual is 'disabled' within the meaning of the Act . . . focuses on her past or present physical condition without regard to mitigation." He claimed that eight of nine federal courts of appeals who had addressed the issue, all three of the executive agencies that had issued regulations or guidelines concerning the ADA, and members of congressional committees who had written reports during preparation of the ADA had taken this view. He also made the point that "if United regards petitioners [the sisters] as unqualified [to be commercial pilots] because they cannot see well without glasses, it seems eminently fair for a court also to use uncorrected vision as the basis for evaluating petitioners' life activity of seeing."

Impact

Unlike the decision in *Bragdon v. Abbott*, the decision in *Sutton v. United Air Lines* suggests that the ADA does not cover people who are presently healthy but have been shown by tests to have a genetic susceptibility that might cause them to become sick or disabled at a later time. The only exception might occur if such people could demonstrate that an employer had excluded them from an entire class of jobs or from any kind of employment because of their genetic predisposition.

---

[1] "Recombinant DNA Advisory Committee: Recombinant DNA and Gene Transfer," NIH fact sheet, p. 1. Available online. URL: http://www4.od.nih.gov/oba/rac/aboutrdagt.htm. Updated March 29, 2000.

[2] "Recombinant DNA Advisory Committee," p. 2.

[3] Mitchel Zoler, quoted in Martin Kenney, *Biotechnology: The University-Industrial Complex* (New Haven: Yale University Press, 1986), p. 256.

[4] Donald Dunner, quoted in Kenney, p. 256.

[5] Peter Farley, quoted in Edward J. Sylvester and Lynn C. Klotz, *The Gene Age: Genetic Engineering and the Next Industrial Revolution* (New York: Scribner, 1983), p. 118.

[6] Office of Technology Assessment (1987), quoted in *John Moore v. Regents of California.*

[7] Dana Hawkins, "Court Declares Right to Genetic Privacy," *U.S. News & World Report*, February 16, 1998, p. 4.

[8] National Action Plan for Breast Cancer, "*Bragdon v. Abbott*: Indications for Asymptomatic Conditions." Available online. URL: http://www.napbc.org/napbc/heredita1.htm. Posted February 1999.

CHAPTER 3

CHRONOLOGY

This chapter presents a chronology of important events relevant to biotechnology, primarily the subset of biotechnology that involves genetic engineering. It also includes key events in the history of genetics, since genetic engineering would have been impossible without the knowledge gained from basic genetic research. It focuses primarily on the period following the invention of genetic engineering in the early 1970s and on events related to the ethical, legal, and social implications of biotechnology, genetic engineering, and human genetics.

ABOUT 10,000 YEARS AGO

- Biotechnology begins along with agriculture. It includes domestication and deliberate breeding of plants and animals, as well as (unknowing) use of microbial processes to make bread, cheese, alcoholic drinks, leather, and other products.

1665

- British scientist Robert Hooke discovers microscopic square bodies in a slice of cork and terms them cells.

1793

- U.S. Congress defines a patentable invention as "any new and useful art, machine, manufacture or composition of matter." Products of nature are not included.

1839

- German biologists Matthias Schleiden and Theodor Schwann propose that cells are the units of which all living things are made.

Chronology

Late 1850s–1860s

- Famed French chemist Louis Pasteur provides a scientific basis for part of traditional biotechnology when he shows that fermentation processes such as those used for millennia to produce wine, beer, buttermilk, and cheese depend on living microorganisms.

1859

- British biologist Charles Darwin publishes *On the Origin of Species*, in which he sets forth the theory of evolution by natural selection. The theory states that the members of a species with inherited characteristics that make them most suited to survive in a particular environment are the most likely to survive and bear young. Over generations, therefore, a change in the environment can cause a change in the predominant characteristics of a species.

1866

- Gregor Mendel, an Austrian monk, publishes a paper describing the basic mathematical rules by which characteristics are inherited. He had worked these out by breeding pea plants in his monastery garden.

1875

- German scientist Walther Flemming discovers that the nuclei (central bodies) of cells contain stringlike bodies that can be stained with dye; these are soon termed chromosomes, or "colored bodies."

1883

- British scientist Francis Galton coins the term *eugenics* (from Greek roots meaning "well born") in a book called *Inquiry into Human Faculty;* he and his followers believe that the human race can be improved by encouraging those with desirable traits to have children and discouraging or preventing those with undesirable traits from doing so.

1900

- Several scientists rediscover Mendel's work, which until this time has been virtually unknown; it now becomes the foundation of genetics.

1910

- American geneticist Thomas Hunt Morgan and coworkers at Columbia University prove that genes are located on chromosomes.

1920s–1930s

- Thirty-four states in the United States pass eugenics laws requiring people with what are thought to be inherited defects (chiefly criminals, the insane, and the "feebleminded") to be forcibly sterilized.

1927

- In *Buck v. Bell*, the U.S. Supreme Court by an 8–1 vote upholds a Virginia eugenics law under which Carrie Buck, a developmentally disabled woman, was forcibly sterilized. Noting that Buck's mother and seven-month-old daughter both appear to be "feebleminded" as well, Justice Oliver Wendell Holmes writes, "Three generations of imbeciles are enough."

1930

- U.S. Plant Patent Act allows breeders to protect plant varieties they have developed by refusing to allow others to reproduce such plants asexually.

1933

- The German government, recently seized by the Nazi Party, passes a eugenics law modeled on those of the United States.

1938

- U.S. Congress passes the Federal Food, Drug, and Cosmetic Act (FFDCA), which gives the Food and Drug Administration (FDA) the right to set tolerances for certain substances, including pesticides, on or in food and feed. Today the FDA shares this regulatory duty with the Environmental Protection Agency (EPA). Herbicide residues on genetically engineered herbicide-tolerant crops and pesticides (such as *Bacillus thuringensis* [Bt] toxin) produced in genetically altered food crops are regulated under the FFDCA.

1941

- George Beadle and Edward L. Tatum of Stanford University show that a single (structural) gene carries the instructions for making a single protein (enzyme).

1942

- In *Skinner v. Oklahoma*, the U.S. Supreme Court strikes down a state law requiring forced sterilization of convicted criminals, saying that procreation is "one of the basic civil rights of man."

Chronology

1944

- Oswald Avery and colleagues at the Rockefeller Institute in New York publish a paper demonstrating that DNA, not protein, carries inherited information.

1947

- U.S. Congress passes the Federal Insecticide, Fungicide, and Rodenticide Act (FIFRA), which regulates the distribution, sale, use, and testing of chemical and biological pesticides. FIFRA has been amended to give the Environmental Protection Agency (EPA) control of these regulations, which cover plants and microorganisms that produce pesticidal substances, such as farm crops genetically modified to make *Bacillus thuringensis* (Bt) toxin.

1952

- Scientists clone frogs by transplanting the nucleus of a frog cell into an unfertilized egg cell with the nucleus removed, a technique called nuclear transfer.
- U.S. Congress revises patent law to state that patentable inventions must be novel, useful, and not obvious "to a person of ordinary skill in the art."

1953

- ***April 25:*** James D. Watson and Francis Crick publish a paper in *Nature* in which they describe the structure of the DNA molecule.
- ***May:*** Watson and Crick publish a second paper offering a theory that explains how DNA's structure allows the molecule to reproduce itself.

1958

- Watson and Crick's theory of how the DNA molecule reproduces itself is confirmed.

1961

- Francis Crick suggests that each "letter" of the genetic code—the unit specifying one amino acid in a protein—consists of three adjoining bases in a DNA molecule.

1961–1966

- Molecular biologists decipher the genetic code, determining which amino acid each of the 64 possible combinations of three bases represents. They also work out the process by which the cell makes the proteins specified by the code.

1970

- U.S. Congress passes the Plant Variety Protection Act, which allows breeders to protect new plant varieties by forbidding others to reproduce the plants sexually.

1971

- Robert Pollack warns Paul Berg about the possible dangers of inserting genes from a cancer-causing virus into bacteria capable of infecting humans.

1972

- ***November:*** Herbert Boyer and Stanley Cohen meet in a Hawaiian delicatessen and begin planning recombinant DNA technology.

1972–1973

- Paul Berg, Herbert Boyer, and Stanley Cohen perform the first experiments in which pieces of DNA from one species are inserted into the genome of another species (recombinant DNA).

1973–1974

- Leading scientists in the field write two letters, published in *Science*, that warn of possible dangers of recombinant DNA experiments and call for a moratorium on some types of experiments.

1974

- Herbert Boyer and Stanley Cohen apply for a patent on their gene-splicing technique and assign all potential royalties from it to their respective universities (University of California, San Francisco, and Stanford University).
- ***October:*** The National Institutes of Health establishes the Recombinant DNA Advisory Committee (RAC) to review the safety of recombinant DNA experiments.

1975

- North Carolina passes a law forbidding employers to discriminate against people with sickle-cell trait. This is the first American law to address genetic discrimination in the workplace.

- ***February 24–27:*** One hundred forty geneticists and molecular biologists meet at Asilomar, California, to draw up safety guidelines for recombinant DNA experiments.
- ***April:*** Congress holds first hearings about safety of recombinant DNA research.

1976

- Robert Swanson and Herbert Boyer found Genentech, the first biotechnology company based on recombinant DNA technology.
- U.S. Congress passes the Toxic Substances Control Act (TSCA), which gives the Environmental Protection Agency (EPA) the authority to review new chemicals before they are introduced into commerce. TSCA is later amended to classify some genetically engineered organisms as "new chemicals" to be regulated under the act.
- ***June:*** National Institutes of Health (NIH) publishes safety guidelines for recombinant DNA experiments, based on the Asilomar guidelines. Britain and Europe adopt similar guidelines.

1977

- Sixteen bills regulating recombinant DNA research are introduced into Congress; none of them passes.

1978

- NIH guidelines are relaxed, reducing containment requirements for many recombinant DNA experiments.
- Genentech makes human insulin in genetically engineered bacteria.
- David Botstein and others develop restriction fragment length polymorphism (RFLP) analysis, which will later be used to locate genes that cause inherited diseases and to identify DNA from particular individuals.

1980

- NIH safety guidelines for recombinant DNA research are relaxed further.
- ***June 16:*** In *Diamond v. Chakrabarty*, a landmark case, the U.S. Supreme Court declares that living things can be patented if humans have altered them.
- ***October:*** Genentech stock is offered to the public for the first time and jumps from $35 to $89 a share in the first few minutes of trading despite the fact that the company has not yet sold any products.

EARLY 1980s

- First transgenic plants and animals produced.
- Monoclonal antibodies are hailed as potential "miracle drugs" for treating cancer and other diseases.

1982

- The FDA grants Genentech permission to sell genetically engineered human insulin.

1983

- Kary Mullis discovers the polymerase chain reaction (PCR), which can be used to duplicate small amounts of DNA many times in a few hours.

1985

- The first transgenic farm animals are created.
- Alec Jeffreys of the University of Leicester, Great Britain, publishes a paper in *Nature* describing what he calls "DNA 'fingerprinting,'" a way to use genetic testing to identify individuals.
- ***October:*** The first U.S. patent for a genetically altered plant is issued.

1986

- First mammals (sheep) are cloned by nuclear transfer technique, using embryonic cells.
- The Toxic Substances Control Act (TSCA) is amended to require an EPA permit for releasing genetically altered organisms into the environment.
- ***June:*** U.S. Office of Science and Technology Policy issues framework for regulation of biotechnology, dividing such regulation among five government agencies.

1987

- U.S. Congress passes the Plant Pest Act, which gives the U.S. Department of Agriculture's Animal and Plant Health Inspection Service (APHIS) the right to regulate any organisms, including those produced by genetic engineering, that are or might be plant pests.
- Colin Pitchfork in Britain and Tommie Lee Andrews in the United States become the first people convicted of crimes on the basis of DNA identification testing.
- ***April:*** The U.S. Patent and Trademark Office extends patentability to animals, cell lines, and genes, including human cells and genes.

- ***April 24:*** Steven Lindow and coworkers oversee spraying of genetically altered bacteria onto strawberry plants in a field in Contra Costa County, California. This is the first deliberate release of genetically engineered organisms into the environment.

1988

- ***April:*** The Harvard Oncomouse, a mouse genetically altered to make it unusually susceptible to cancer and thus useful in cancer research and carcinogen testing, becomes the first genetically engineered animal to be patented.

1989

- ***May 22:*** Steven Rosenberg and coworkers at the National Institutes of Health carry out the first successful insertion of foreign genes into a human being. The genes are intended only as markers and have no effect on health.
- Virginia opens the first state database of DNA identification profiles taken from convicted criminals.

1990

- The Human Genome Project, which aims to sequence all human genes by a few years into the 21st century, begins.
- U.S. Congress passes the Americans with Disabilities Act (ADA), which might ban discrimination against healthy people with inherited defects revealed by genetic tests.
- ***July:*** The California Supreme Court rules in *John Moore v. Regents of University of California* that people do not retain ownership of their cells or tissues once these have been removed from their bodies.
- ***September 14:*** W. French Anderson and colleagues from the NIH administer the first successful gene therapy to four-year-old Ashanthi deSilva.

1991

- The FBI establishes guidelines for forensic DNA testing.
- James Watson, head of the Human Genome Project, earmarks 3 percent of the project's budget for investigation of its ethical, legal, and social implications.
- ***September:*** The National Biodiversity Institute (INBio) of Costa Rica promises to provide samples of the country's wild plants, microbes, and insects to drug giant Merck in return for a large research and sampling

budget, training, and royalties from any resulting products; some of the money will be used to preserve the country's national parks.

- New York lawyers Barry Scheck and Peter Neufeld establish the Innocence Project, which seeks to obtain DNA testing for convicted criminals who say that they are innocent and that such testing could clear them.

1992

- U.S. Department of Defense begins requiring all military personnel to give DNA samples, which will be retained in a databank to help identify remains of soldiers killed in action.
- ***May:*** The FDA states that genetically altered foods do not have to receive special approval or be labeled as such if they are nutritionally the same as their natural equivalents and contain no new substances that might cause an allergic reaction.

1993

- The FDA approves recombinant bovine growth hormone (rBGH) for sale; it is the first genetically engineered animal hormone approved for sale in the United States.
- ***March:*** The USDA streamlines its requirements for testing new varieties of genetically engineered corn, cotton, soybeans, potatoes, tomatoes, and tobacco to eliminate a former requirement for an environmental review before testing.
- ***October:*** Robert Stillman and Jerry Hall of George Washington University Medical Center in Washington, D.C., clone early-stage human embryos. The embryos, already due to be discarded by a fertility clinic, are not allowed to develop.

1994

- The Flavr Savr tomato becomes the first genetically engineered food to go on sale.
- ***September 13:*** Congress passes the DNA Identification Act, which gives the FBI national responsibility for training, funding, and proficiency testing of laboratories doing forensic DNA profiling and authorizes it to establish a national database of DNA information from crime scenes and convicted criminals.
- The Equal Employment Opportunity Commission (EEOC) rules that denying employment to people who are healthy but have inherited defects revealed by genetic tests violates the Americans with Disabilities Act.

- The Environmental Protection Agency (EPA) decides to regulate plants genetically engineered to produce their own pesticides as if the plants were chemical pesticides.
- The USDA and W. R. Grace, a multinational corporation, obtain a patent for using the oil of the neem tree, native to India, against fungi on plants. People in India have used the oil in this way for millennia, and activists call the patent an example of biopiracy.

1995

- The World Trade Organization's (WTO) Trade Related Intellectual Property Rights Agreement (TRIPS) goes into effect. This agreement requires WTO member countries to "harmonize" their patent laws with those of the United States and other industrialized countries, supposedly in order to facilitate the transfer of technology to those countries. Critics say its real purpose is to protect the rights and profits of multinational corporations.
- Britain establishes the Criminal Justice DNA Database, the world's first database of DNA identity profiles of convicted criminals. It covers all crimes that result in imprisonment.
- ***October:*** Former football star O. J. Simpson is acquitted of the murder of his wife and an acquaintance, Ronald Goldman. The verdict comes despite DNA evidence linking Simpson to the crimes, after his defense lawyers show that samples from the crime scene were mishandled and may have been accidentally or deliberately contaminated with Simpson's blood before testing.

1996

- Genetically altered crops, including those that resist herbicides and those that produce their own insecticide, are sold commercially and planted on a large scale in the United States for the first time.
- Several studies reveal evidence of discrimination in insurance and employment on the basis of the results of genetic tests.
- ***April:*** U.S. Marines John C. Mayfield III and Joseph Vlakovsky are court-martialed and found guilty of disobeying a direct order because they refused to give DNA samples for storage in a military database, saying that doing so invaded their genetic privacy.
- ***August 21:*** U.S. Congress passes the Health Insurance Portability and Accountability Act, which includes a provision forbidding health insurers issuing group plans to deny coverage to healthy people because of preexisting genetic conditions.

1997

- The British government's Human Genetics Commission recommends a two-year moratorium on disclosure of genetic testing results to insurers. The moratorium is later extended for another five years.
- ***January:*** Deputy Attorney General Jamie Gorelick publicly admits that the FBI crime laboratory has a "serious set of problems" in its handling and testing of forensic DNA samples.
- ***February 27:*** Ian Wilmut and coworkers at the Roslin Institute in Scotland publish an article in *Nature* announcing that they have cloned a sheep (Dolly) from a mature udder cell taken from a six-year-old (adult) ewe.
- ***March:*** Senator Christopher Bond and Representative Vernon Ehlers introduce bills that will ban human cloning.
- ***March 4:*** Reacting to fears stirred by the news about Dolly, President Bill Clinton announces a ban on the use of federal funds for research on cloning of human beings.
- ***March 5:*** The first Congressional hearing on human cloning takes place before the House Science Committee's Technology Subcommittee.
- ***April 11:*** U.S. Congress amends the Toxic Substances Control Act (TSCA) to tailor the general screening program for microbial products of biotechnology to meet the special requirements of microorganisms used commercially for such purposes as production of industrial chemicals and breakdown of chemical pollutants in the environment (bioremediation).
- ***June:*** A report from the National Bioethics Advisory Commission recommends a legislative moratorium on human cloning for three to five years.
- ***November:*** The European Union begins requiring labeling of all genetically modified foods.
- ***November:*** The General Conference of the United Nations Educational, Scientific and Cultural Organization (UNESCO) adopts the Universal Declaration on the Human Genome and Human Rights, which asserts, among other things, that "the human genome in its natural state shall not give rise to financial gains" and affirms individuals' rights to have genetic tests or research performed on them only with consent, to be free of discrimination based on genetics, and to have their genetic information kept private.

1998

- The European Union places an informal moratorium on approval of new genetically modified (GM) crops and importation of any food products that might contain unapproved GM materials.

- Responding to a 1997 suit from a group of California women on death row (*Alfaro v. Terhune*), a Sacramento superior court judge grants an injunction to prevent DNA samples being taken from death row inmates pursuant to a California law. The judge says that the sampling procedure is flawed, but he rules that the law itself is constitutional.
- ***January:*** California becomes the first state to outlaw human reproductive cloning.
- ***February:*** In *Norman-Bloodsaw v. Lawrence Berkeley Laboratory*, the U.S. Court of Appeals for the 9th Circuit rules that Lawrence Berkeley Laboratory's genetic and other testing of employees' blood and urine samples without their informed consent was unconstitutional.
- ***June:*** In *Bragdon v. Abbott*, the Supreme Court rules that the Americans with Disabilities Act (ADA) protects a healthy woman who is HIV positive, leading to speculation that it may also apply to healthy people with genetic problems revealed by testing.
- ***July 6:*** After a decade of heated discussion and intense lobbying, the European Parliament and Council approve Directive 98/44, on the legal protection of biotechnological inventions. It permits patenting of plants but, on ethical grounds, bars patents on processes for cloning or modifying the genetic identity of human beings, use of human embryos for industrial purposes, and processes for modifying the genetic identity of animals that may cause them suffering unless such processes are likely to produce substantial medical benefits for humans. It allows patenting of gene sequences only for particular industrial applications. It states that member countries may reject patents that they feel violate moral or ethical standards.
- ***October:*** The FBI opens its National DNA Index System (NDIS), a national database that will store information from DNA samples taken from convicted murderers and sex offenders by the states, and CODIS, a program to coordinate information from state and national DNA databases.
- ***November:*** Geron Corp. and James A. Thomson of the University of Wisconsin announce that they have isolated and cultivated human embryonic stem cells, which potentially can provide tissues for transplantation. Further research on these cells may involve the use of cloned human embryos.

1999

- Ingo Potrykus, Peter Beyer, and other researchers create "golden rice," which contains added daffodil and bacterial genes that make it produce beta-carotene, a building block of vitamin A. Biotechnology supporters tout this genetically engineered crop as a cure or preventive for vitamin A deficiency, which affects up to 140 million children each year and makes about 500,000 of them go blind.

- After 20 years of disappointing results, monoclonal antibody drugs finally begin to be marketed as treatments for certain types of cancer.
- ***May:*** A Cornell University study suggests that pollen from corn crops genetically altered to produce an insecticide may land on milkweed plants and poison monarch butterfly larvae.
- ***June 22:*** In a Supreme Court decision on *Sutton v. United Air Lines*, Justice Sandra Day O'Connor writes that the ADA protects only those who are presently disabled, dimming hopes that this law can be applied to healthy people with genetic predispositions to disease.
- ***September 17:*** Jesse Gelsinger, an 18-year-old Arizona man with a rare inherited disease, dies after taking part in a gene therapy experiment. His is the first death officially attributed to gene therapy.
- ***November 2:*** Researchers and drug companies admit having concealed from the NIH six deaths during a different gene therapy experiment. They claim that the therapy did not cause the deaths.

2000

- ***January 21:*** Reporting on inspections conducted after the death of Jesse Gelsinger, the FDA sharply criticizes procedures in his and other gene therapy trials and suspends some of them.
- ***January 29:*** The compromise Cartagena Biosafety Protocol, which urges a "precautionary approach" to genetically modified crops, is completed and signed by representatives of more than 130 countries in Montreal, Quebec.
- ***February:*** Massachusetts-based Aqua Bounty Farms applies for federal approval for the first genetically modified animal intended for human consumption, a transgenic salmon that grows twice as fast as normal salmon.
- ***February:*** President Bill Clinton signs Executive Order 13145, which prohibits genetic discrimination in federal employment.
- ***February:*** Nearly 250 concerned environmentalists and other leaders sign an open letter warning that human germ-line engineering "is an unneeded technology that poses horrific risks" and urging that it be banned.
- ***April:*** French scientists report that gene therapy has apparently cured several children with severe combined immune deficiency.
- ***May 10:*** The European Patent Office revokes the USDA–W. R. Grace patent on using neem oil against fungi on the grounds that it lacks novelty and inventiveness.
- ***June 26:*** Teams of scientists from the private company Celera Genomics and the government-funded National Human Genome Research Insti-

tute announce that they have finished a rough draft of the code of the complete human genome.

- ***August 29:*** A woman in Denver, Colorado, gives birth to a boy who was chosen by preimplantation genetic diagnosis not only to be free of an inherited disease that affected his older sister but to be of the same tissue type as the sister so that a stem cell transplant from him could cure her. Some people criticize this move because it seems too close to conceiving a child simply to produce "spare parts."
- ***September 18:*** The environmental group Friends of the Earth announces that Taco Bell taco shells contain traces of a type of genetically modified corn called StarLink, which had been approved for animal feed but not human food. StarLink is eventually found to have contaminated more than 430 million bushels of corn, and more than 300 types of corn-containing food products have to be recalled.
- ***September 19:*** Jesse Gelsinger's father sues the researchers and institutions responsible for the gene therapy that caused his son's death.
- ***October 30:*** A group of families with Canavan syndrome, a rare inherited condition, file suit against Miami Children's Hospital for breach of contract and violation of informed consent requirements. They claim that the hospital used donated material from their bodies to isolate the gene that causes the disease and then, instead of making a diagnostic test for the illness available to all as it had implied when collecting the samples, it took out a patent on the gene that made the test restrictively expensive.
- ***November:*** The Jesse Gelsinger lawsuit is settled out of court.

2001

- A fertility clinic announces that about 30 "normal, healthy" babies worldwide have been created by a technique that combines the nucleus of an egg from one woman with cytoplasm from the egg of another and sperm from a man, resulting in children whose cells contain DNA from all three people. This technique, done to avoid diseases caused by defective mitochondrial DNA in the nucleus donor, in effect alters germline genes.
- ***January:*** Scientists announce creation of the first genetically modified primate, a rhesus macaque monkey named ANDi (short for "inserted DNA" backward), whose cells contain a jellyfish gene used as a marker.
- ***January:*** Australian researchers say they accidentally found a way to modify the genes of a mouse virus that made it much more deadly, adding to fears that genetic engineering could be used to make exceptionally dangerous bioterrorism weapons.

- ***January 5:*** The U.S. Patent and Trademark Office issues final revised guidelines on the utility requirements for patenting human genetic sequences, affirming the "specific, substantial, and credible" standard it had established in 1999 but reemphasizing that an unaltered human gene could be patented as long as it had been isolated and purified.
- ***January 22:*** Britain legalizes cloning of human embryos for stem cell research.
- ***February 9:*** The federal Equal Employment Opportunity Commission (EEOC) files suit against the Burlington Northern Santa Fe Railroad, claiming that it violated the Americans with Disabilities Act (ADA) by making secret genetic tests of employees who filed injury claims for carpal tunnel syndrome.
- ***spring:*** The University of Hawaii tentatively agrees to demands by an organization of families with the rare hereditary condition pseudoxanthoma elasticum, promising the families a share in control of and royalties from any patent on the gene that causes the disease in return for their donation of biological material to the university's researchers.
- ***April 16:*** Federal district court judge Albert V. Bryan, Jr., rules that denying convict James Harvey a chance to have DNA samples used in his trial retested, a move that Harvey claims would prove his innocence, violates the right to due process of law guaranteed by the Fourteenth and Fifteenth Amendments.
- ***April 18:*** The Burlington Northern Santa Fe Railroad agrees to an out-of-court settlement of the EEOC suit, in which it admits no wrongdoing but promises to stop all genetic tests of workers. It later also agrees to pay the tested employees $2.2 million.
- ***August:*** The Centers for Disease Control and Prevention reports that it was unable to detect antibodies to the bacterial protein Cry9C in any of the people who claimed to have suffered allergic reactions to StarLink-contaminated corn, thus making it unlikely—though not impossible, the agency admits—that the corn had caused the people's health problems.
- ***August 9:*** U.S. President George W. Bush announces that henceforth, federal funding can be used for research only on embryonic stem cell lines that already exist on this date. The ruling does not affect privately funded research.
- ***fall:*** New studies suggest that milkweed plants on which monarch butterfly caterpillars feed are unlikely to receive enough pollen from genetically modified, insecticide-containing corn to endanger the caterpillars.
- ***fall:*** Kenyan farmers harvest the first crop of yams, a food widely eaten in Africa, that have been genetically engineered to resist a virus that normally destroys up to 80 percent of the crop.

- ***October 9:*** The European Court of Justice rejects the claim of the Netherlands, Italy, and Norway that living things and human biological material should not be patentable and reaffirms the demand in European Union (EU) Directive 98/44 that the patent laws of all EU member countries be "harmonized."
- ***November:*** Advanced Cell Technology, a biotechnology company in Worcester, Massachusetts, announces that it has cloned the first human embryos for use in stem cell research. The embryos stopped development far sooner than the stage at which stem cells could be harvested, however, and many scientists question the importance of the company's achievement.
- ***November 29:*** Ignacio Chapela and David Quist of the University of California, Berkeley, publish an article in *Nature* claiming that they have found genes from bioengineered corn in Mexican corn crops, even though that country does not allow the planting of genetically engineered corn. This finding suggests that gene flow, which could lead to environmental damage, is taking place.

2002

- The British government's Human Fertilisation and Embryology Authority allows a couple to use preimplantation genetic diagnosis to select an embryo with a tissue type compatible with that of a sick older sibling because the embryo also needs to be selected to be free of a genetic disease, but it denies permission for another couple to do the same when determination of tissue type is the only reason for testing.
- A deaf lesbian couple deliberately chooses a sperm donor with a strong strain of inherited deafness in his family to ensure that their child will be born deaf and thus will share fully in the deaf culture, which they value highly. Many people find this a controversial use of genetic selection.
- ***January:*** A three-judge panel of the Fourth Circuit Court of Appeals reverses district court judge Albert Bryan's decision in *Harvey v. Horan*, ruling that felons do not have a constitutional right to demand DNA testing after their conviction.
- ***February:*** Scientists at Texas A&M University and a biotechnology firm called Genetic Savings & Clone announce that they have cloned a cat, raising some people's hopes of "reincarnating" beloved pets who have died.
- ***February:*** James Harvey is granted access to the DNA sample he requested, thanks to a new state law in Virginia.
- ***April:*** Severino Antinori, an Italian gynecologist, claims that a woman under his care is eight weeks pregnant with a cloned embryo. No later evidence of the birth of a clone appears.

- ***April 11:*** *Nature* takes the unprecedented step of apologizing for publishing Ignacio Chapela and David Quist's paper claiming contamination of Mexican corn, saying that the authors' evidence is not sufficient to justify their conclusions.
- ***May:*** Reviewing *Alfaro v. Terhune*, the Third District Court of Appeals unanimously rejects the argument of California death row inmates that requiring them to give blood and saliva samples for the state's forensic DNA database violates their right to privacy and their Fourth Amendment right to be free of unreasonable searches and seizures.
- ***July:*** Scientists make polio viruses "from scratch," using gene sequences from a mail-order supplier and instructions downloaded from the Internet, suggesting that bioterrorists might also be able to make their own viruses.
- ***July:*** The President's Council on Bioethics issues a report on human cloning in which all 17 members recommend a permanent ban on reproductive cloning. Seven members support research or therapeutic cloning under careful regulation, but a 10-member majority votes for a four-year moratorium on research cloning so that additional public discussion of this practice can take place.
- ***August:*** In spite of facing probable famine, the African countries of Zambia, Zimbabwe, Mozambique, and Malawi refuse to accept U.S. corn from the UN World Food Programme unless it is milled first. They fear that farmers might plant the corn, which could be genetically modified, and thereby ruin the countries' chances of selling corn to Europe, which forbids any importation of genetically modified food.
- ***August:*** The California Supreme Court declines to review *Alfaro v. Terhune*, allowing the appeals court ruling that rejected the death row inmates' suit against DNA testing to stand.
- ***August:*** On the basis of tests matching his DNA to that in semen found on the body of five-year-old Angela Bugay, a jury convicts Larry Graham, a California man, of raping and murdering the girl in 1983. His case is cited as an example of the solution of a long-past, "cold case" crime through DNA testing.
- ***September:*** French scientists report that a child whom gene therapy had apparently cured of severe inherited immune deficiency has developed a leukemia-like blood condition. The overproduced cells in the child's blood contain the inserted gene, which suggests that the therapy somehow caused the cancer.
- ***September:*** California governor Gray Davis signs a new law that allows the taking of DNA samples from convicted felons by force. Taking of samples from death row inmates in the state resumes.

- ***October:*** The Food and Drug Administration (FDA) announces that it will regulate genetically altered animals and their products as drugs, providing a review that is strict but not open to the public.
- ***November:*** An ounce of corn genetically engineered to contain a vaccine, made by a company called ProdiGene, is found mixed with soybeans intended to become human food. The material is removed before it enters the food supply, and the USDA orders Prodigene to pay a $250,000 fine for not following safety procedures.
- ***December:*** In rejecting a patent application for a genetically engineered mouse, the Supreme Court of Canada declares that higher life-forms (multicellular differentiated organisms, including plants and animals) cannot be patented in that country.
- ***December 27:*** A flying saucer cult called the Raelians announces that it has produced a cloned human baby named Eve. The Raelians fail to provide convincing evidence of their claim, however, and most experts later conclude that it is false.

2003

- ***January:*** The FDA temporarily halts some gene therapy trials in the United States after learning that a second French child cured of severe inherited immune deficiency by a gene treatment has developed a form of leukemia.
- ***February:*** Dolly, the famous cloned sheep, is euthanized at the age of eight years because she has developed a severe lung disease.
- ***February 27:*** The U.S. House of Representatives passes a bill that bans all forms of human cloning, including research cloning.
- ***March:*** The British government announces that police will be able to keep DNA profiles indefinitely from everyone who has donated specimens, whether or not the people have been accused of a crime.
- ***spring:*** Food industry and environmental groups call for a moratorium on planting of experimental crops genetically engineered to produce drugs and industrial chemicals. The USDA announces that it is tightening regulation on such crops.
- ***April:*** A district court judge accepts an agreement by Aventis and other companies that created and distributed StarLink corn to pay $110 million to farmers in an out-of-court settlement of a class action suit.
- ***April:*** Scientists clone a banteng, an endangered cattlelike animal from Java. This is the first clone of an endangered species to survive more than a few days beyond birth.
- ***April:*** The Justice Department inspector general, an independent watchdog, investigates the FBI's DNA-testing laboratory in the wake of

a discovery that a technician there had failed to follow proper DNA-testing procedure for two years before being detected.

- ***April 14:*** The International Human Genome Sequencing Consortium announces that the Human Genome Project is completed, more than two years ahead of schedule.
- ***May 13:*** The United States sues the European Union, claiming that proposed new regulations restricting genetically engineered foods are not based on science and violate World Trade Organization rules.
- ***June:*** A genetically engineered vaccine for pneumonic plague is announced to be very effective in tests on mice.
- ***July 2:*** The European Parliament approves new regulations that require importing countries to provide traceability of food products "from farm to fork" and verify that foods not labeled as GM-containing have no more than 0.9 percent accidental GM contamination.
- ***September 9:*** In a case involving Italy, the European Court of Justice rules that individual EU member states may temporarily ban genetically engineered foods approved by other member states but stresses that such bans must be scientifically justified.
- ***September 11:*** The Cartagena Biosafety Protocol, the United Nations treaty governing international movement of genetically modified organisms, goes into effect.
- ***October 2:*** A three-judge panel of the Ninth U.S. Circuit Court of Appeals votes 2-1 to strike down a federal law that requires parolees not suspected of new crimes to donate blood samples for the national DNA database, stating that it violates the parolees' constitutional protection against unreasonable search and seizure.
- ***October 14:*** The U.S. Senate unanimously passes a bill that would prevent companies from denying employment or health insurance coverage on the basis of genetic information.
- ***October 31:*** The U.S. Food and Drug Administration issues a draft report stating that food made from cloned animals and their offspring is no more dangerous than comparable food from noncloned animals.
- ***November 7:*** Controversial new European Union regulations on labeling and traceability of genetically modified organisms go into effect.

2004

- ***January:*** Bowing to a request by the Justice Department, the Ninth Circuit Court of Appeals agreed to rehear en banc (with all judges) the case that caused a three-judge panel from the court to overturn a federal law requiring parolees to give blood samples to the national DNA database.

- ***March:*** Two of the three full-time scientists on the President's Council on Bioethics, one of whom had just been removed from the committee, claim that the council ignored or distorted scientific evidence when writing recent reports on stem cell research and on research to combat age-related disease.
- ***March:*** A Mexican government study reports discovery of altered genes in 7.6 percent of sampled corn plants in Oaxaca state, supporting the disputed earlier claims of David Quist and Ignacio Chapela that gene flow from engineered to natural corn crops is occurring in that country.

CHAPTER 4

BIOGRAPHICAL LISTING

This chapter offers brief biographical information on people who have played major roles in the development of biotechnology and genetic engineering or in events relating to the ethical, legal, and social implications of these fields. Some pioneers of genetics and molecular biology are also included, but the list focuses primarily on people who have been active since the development of genetic engineering in the early 1970s.

Sidney Abbott, an HIV-positive woman with no symptoms of AIDS. She sued her dentist, Randon Bragdon, for refusing to treat her, saying that his refusal violated her rights under the Americans with Disabilities Act (ADA). In June 1998 the Supreme Court confirmed her right to sue, leading to hope that the ADA might also cover healthy people with a genetic predisposition to disease.

Maria Alfaro (Rosie Alfaro), alphabetically first among a group of women death row inmates who in 1997 sued the California Department of Corrections to prevent their being required to give DNA samples to the state's criminal database. A California Superior Court judge granted them an injunction in 1998, but in May 2002 the Third District Court of Appeals unanimously ruled against their claims that the requirement violated their constitutional right to privacy and to be free from unreasonable searches and seizures. In August 2002 the California Supreme Court declined to review the case *(Alfaro v. Terhune)*, letting the appeals court's ruling stand.

W. French Anderson, pioneer in gene therapy. He headed the National Institutes of Health (NIH) team that made the first successful use of altered genes to treat a human disease in September 1990. He is currently at the University of Southern California.

Tommie Lee Andrews, the first American convicted of a criminal charge primarily through DNA identification testing. Andrews, a 24-year-old warehouse worker, was convicted of the rape and beating of several women in Orlando, Florida, in late 1987 and early 1988 after the DNA profile from a sample of his blood matched that of semen found on one of the victims.

Severino Antinori, Italian gynecologist. In April 2002, Antinori announced that a woman under his care was eight weeks pregnant with a cloned embryo, but no actual child was ever produced.

Oswald Avery, Canadian-born bacteriologist who proved that DNA rather than proteins carries inherited information. In 1944, Avery and his colleagues at the Rockefeller Institute in New York published a paper showing that a form of *Pneumococcus* bacteria incapable of causing disease became able to cause illness after it took up DNA from a disease-causing form of the same bacteria. The change was passed on to the bacteria's descendants.

Paul Berg, Stanford University biochemist sometimes called "the father of genetic engineering." Berg did the first recombinant DNA experiments, transferring genes between two types of viruses, in the winter of 1972–73. Berg was also one of the first scientists to question the safety of recombinant DNA experiments.

Peter Beyer, scientist at the Swiss Federal Institute of Technology. He and Ingo Potrykus led the team that created "golden rice" in 1999. This rice contains daffodil and bacterial genes that allow it to make beta-carotene, a building block of vitamin A. Biotechnology supporters claim that eating it can prevent vitamin A deficiency, which affects 100 to 140 million children a year and makes about 500,000 of them go blind.

Larry Bohlen, environmental activist and director of health and environment programs for Friends of the Earth. In September 2000 he took corn products from a local market to a laboratory and had them tested. The laboratory found traces of StarLink, a type of genetically engineered corn that had been approved for animal feed but not human food, in Taco Bell taco shells. StarLink was later found to have contaminated a wide variety of other corn products as well, causing a major controversy.

Herbert Boyer, a biochemist. In 1973, while at the University of California at San Francisco, he worked with Stanley Cohen of Stanford to develop the first practical technique for transferring genes between organisms. Boyer and venture capitalist Robert Swanson cofounded Genentech, the first business built on recombinant DNA technology, in 1976.

Albert V. Bryan, Jr., U.S. District Court judge in Virginia. In April 2001 he ruled in *Harvey v. Horan* that felons had a constitutional due process right to obtain DNA samples for testing that might prove their innocence. The U.S. Fourth Circuit Court of Appeals reversed his ruling in January 2002.

Carrie Buck, an apparently developmentally disabled woman in an institution who appealed a decision to forcibly sterilize her under a Virginia eugenics law. The U.S. Supreme Court upheld the law by an 8-1 vote in 1927. Noting that Buck's mother and seven-month-old daughter as well as Buck herself seemed to be "feebleminded," Justice Oliver Wendell Holmes wrote in the majority opinion on the case that "three generations of imbeciles are enough." Buck may in fact have had normal intelligence.

Ananda Chakrabarty, General Electric Corporation scientist. He obtained the first U.S. patent on a living thing. His application for a patent on a genetically engineered bacterium that digested petroleum was rejected at first by the Patent and Trademark Office, but in 1980 the Supreme Court ruled by a 5-to-4 vote that Congress had intended "anything under the sun that is made by man" to be patentable, including altered organisms such as the one Chakrabarty had invented.

Ignacio Chapela, researcher at the University of California, Berkeley. His paper, coauthored with David Quist, in the November 29, 2001, issue of *Nature* caused widespread controversy by claiming that engineered genes had appeared in native corn crops in Mexico. On April 11, 2002, *Nature* took the unprecedented step of apologizing for publishing the paper, saying that evidence did not justify the authors' conclusions. Chapela, Quist, and some other scientists stand by their report, however, and later reports by other scientists seem to confirm it.

Stanley Cohen, Stanford University geneticist. With Herbert Boyer, he invented the first practical method of transferring genes between organisms in 1973. Cohen had studied plasmids, small rings of DNA that bacteria sometimes use to transfer genetic information, and he and Boyer worked out a method of combining plasmids from two types of bacteria.

Francis Collins, a molecular biologist. He currently directs the National Institutes of Health's National Human Genome Research Institute and the Human Genome Project.

Francis Crick, British molecular biologist. With James Watson, he worked out the structure of the DNA molecule at Cambridge University and showed how this structure allowed the molecule to reproduce itself. Watson, Crick, and Maurice Wilkins shared a Nobel Prize for this work in 1962. In the early 1960s, Crick set out a theory explaining how inherited information was coded in DNA and how the cell translated this information into proteins. His ideas were later shown to be correct.

Charles Darwin, author of the theory of evolution by natural selection, which he set forth in *On the Origin of Species* (1859). The theory states that the members of a species with inherited characteristics that make them most suited to survive in a particular environment will be the most likely to survive and bear young. Over generations, therefore, a change in the environment can cause a change in the predominant characteristics of a species.

Ashanthi deSilva was four years old in September 1990, when she made history as the first person to receive altered genes as a treatment for disease. Ashanthi had inherited a rare condition called ADA (adenosine deaminase) deficiency, which left her without a functioning immune system. After several treatments in which her own blood cells were reinjected after being given normal ADA genes in the laboratory, she became healthy enough to lead an essentially normal life.

Walther Flemming, a German biologist. In 1875, he discovered chromosomes ("colored bodies") in the nuclei or central bodies of cells and showed how these changed as cells reproduced.

Francis Galton, a cousin of Charles Darwin. He founded the pseudoscience of eugenics, which held that personality characteristics such as intelligence or laziness are inherited and that the human race can be improved by encouraging people with desirable characteristics to have children and discouraging reproduction in those with undesirable characteristics. Some of Galton's supporters advocated sterilization by force. Eugenics has fallen into disrepute, but Galton should also be remembered as a founder of biostatistics (application of statistics to animal populations) and of identification by fingerprinting.

Jesse Gelsinger, an 18-year-old Arizona man with a rare genetic disease. He died on September 17, 1999, when his immune system reacted violently to viruses injected into him as part of an experimental gene therapy treatment. His death was the first attributed to gene therapy. It resulted in suspension or cancellation of a number of gene therapy trials in the United States and a considerable tightening of the regulations governing human trials.

David Golde, physician at the University of California, Los Angeles (UCLA) Medical Center, who was a defendant in a suit concerning ownership rights to tissue that was made into a commercial cell line. John Moore, whom Golde had treated for leukemia, sued Golde and others in 1984 because they had developed a lucrative cell line from Moore's spleen (removed as part of his cancer treatment) without obtaining Moore's permission or offering him any recompense. The California Supreme Court ruled in 1990 that Moore had no ownership rights to his tissue but also stated that Golde had violated his "fiduciary duty" to Moore by not telling him about his commercial plans.

Larry Graham, a California man, was convicted in August 2002 of raping and murdering a five-year-old girl 19 years before. Testing that matched his DNA with that in semen found on the girl's body was crucial in the conviction, and his case was cited as an example of the "cold cases" that could be solved, years after the crimes were committed, through DNA testing and databases.

James Harvey, a Virginia man convicted of rape and forcible sodomy in 1990. He protested his innocence and, with the help of the New York–based Innocence Project, demanded access to the rape kit used as evidence in his trial so that improved DNA testing methods could be applied to it. When the prosecutor, Robert F. Horan, Jr., refused to hand over the kit, Harvey and his lawyers sued Horan in July 2000, claiming violation of 42 U.S.C. 1983, a civil rights statute. A District Court judge stated in April 2001 that Horan's refusal violated Harvey's due process rights under the Fourteenth and Fifteenth Amendments, but in January 2002 the U.S. Fourth Circuit Court of Appeals reversed the lower court's decision, saying that convicts did not have a constitutional right to demand DNA testing after their conviction. In February, however, another judge, citing a new Virginia state law, ordered that Harvey be given the rape kit. The DNA test, when finally performed, confirmed Harvey's guilt.

Kimberly Hinton, one of two twin sisters who applied for the job of global airline pilot with United Airlines in the 1990s (the other was Karen Sutton). The airline rejected her because she failed to meet its requirement for uncorrected vision, since she was extremely nearsighted without glasses, although her vision with corrective lenses was 20/20 or better. She and her sister sued the airline for violation of the Americans with Disabilities Act (ADA), but the U.S. Supreme Court rejected their suit in June 1999 because, as Sandra Day O'Connor wrote in the Court's majority opinion, the ADA requires someone to be "presently—not potentially or hypothetically—substantially limited" in a major life activity. The court did not consider the women disabled because their disability could be corrected. This decision suggested that people with a genetic predisposition to disease but no actual illness also would not be covered by the ADA.

Oliver Wendell Holmes, Supreme Court Justice from 1902 to 1932. In the 1927 case *Buck v. Bell,* Holmes wrote the majority opinion that supported a Virginia eugenics law under which a developmentally disabled woman, Carrie Buck, was forcibly sterilized. Noting that not only Buck but her mother and her seven-month-old daughter appeared to be of subnormal intelligence, Holmes wrote that "three generations of imbeciles are enough."

Leroy Hood, a pioneer inventor of machines that handle DNA automatically. In 1983 he and coworkers at the California Institute of Technology invented a machine that could assemble short stretches of DNA of known sequence. They invented one that could automatically determine the sequence of stretches of DNA in 1986. Hood's machines have helped make gargantuan tasks like the Human Genome Project possible.

Biographical Listing

Robert Hooke, British scientist, inventor, and philosopher. In addition to many achievements in chemistry and physics, Hooke invented or greatly improved the compound microscope. In 1665, using this invention, he observed tiny, square structures in a slice of cork and dubbed them cells. In fact, what he saw were merely the walls that were all that remained of these basic units of living matter.

Alec Jeffreys, of the University of Leicester in Great Britain, invented the technique of "DNA 'fingerprinting,'" as he called it, in 1985. Based on the observation that certain stretches of DNA contain short repeated base sequences that vary considerably in length from person to person, Jeffreys's technique was used first in immigration and paternity cases. It began to be used for identification of criminals in 1987.

Leon R. Kass, University of Chicago bioethicist, heads the President's Council on Bioethics. He is also a fellow at the American Enterprise Institute. Kass disapproves strongly of human cloning for any purpose and of alteration of human germ-line genes.

Steven Lindow, a plant pathologist at the University of California at Berkeley. He oversaw the first release of genetically altered organisms (bacteria) into the environment in 1987. His "ice-minus" bacteria protected crops against frost damage.

Victor McKusick, director of medical genetics at Johns Hopkins Medical School in Baltimore, called "the father of genetic medicine." He chaired the National Research Council (NRC) committee that produced a controversial report on forensic DNA testing in 1992.

John C. Mayfield III, a 21-year-old lance corporal in the U.S. Marines in 1995. He and Corporal Joseph Vlakovsky refused to give samples for DNA testing and archiving, which had been required of all U.S. military personnel since 1992. Mayfield and Vlakovsky said that the DNA database, which is supposed to help identify the remains of soldiers killed in action, violated their right to privacy. A court-martial in April 1996 convicted the two of refusing to obey a direct order.

Gregor Mendel, an Austrian monk. He bred pea plants in his monastery garden to work out the basic rules by which characteristics are inherited. He published a paper describing his work in 1866, but it remained virtually unknown until several scientists rediscovered it in 1900. Mendel's work became the foundation of genetics and provided the mechanism for evolution by natural selection, first described by Mendel's contemporary Charles Darwin.

John Moore, a Seattle businessman. He sued his doctor, David Golde, the University of California, and several drug companies in 1984 for having made a profitable laboratory cell line from his spleen, which had been removed as a cancer treatment, without consulting him. A lower court supported him, but in 1990 the California Supreme Court ruled that Moore had no ownership claims on his cells once they had been removed from his body.

Thomas Hunt Morgan, pioneer geneticist, worked at Columbia University in New York (1904–28) and later at the California Institute of Technology (1928–45). At Columbia, he and his coworkers and students drew on experiments with fruit flies to relate changes in chromosomes to the inheritance of particular characteristics, thus providing a physical basis for the patterns of trait transmission that Gregor Mendel had observed. Morgan won a Nobel Prize for his work in 1933.

Kary Mullis, discoverer of the polymerase chain reaction (PCR) in 1983, when he was working for Cetus Corporation in Emeryville, California. Using an enzyme called polymerase, PCR repeatedly duplicates DNA, allowing tiny samples to be increased rapidly to a quantity suitable for analysis. It has proved useful in everything from genome sequencing to testing of DNA samples from crime scenes. Mullis won a Nobel Prize for his discovery in 1994.

Peter Neufeld, a New York defense attorney. He and his partner, Barry Scheck, have become famous for acting in criminal cases involving identification by DNA testing and have appeared in a number of trials, including the O. J. Simpson murder trial in 1995. In 1991 they founded the Innocence Project, which has used DNA testing to show that certain convicted criminals were actually not guilty.

Marya Norman-Bloodsaw, a worker at Lawrence Berkeley Laboratory, a California research facility run by the Department of Energy. In 1995, she and some of her coworkers found out that the laboratory had run several tests, including a genetic test, on samples of their blood and urine without their knowledge or consent. They sued for violation of their rights under the Fourth Amendment and the 1964 Civil Rights Act, and the U.S. Court of Appeals for the Ninth Circuit upheld them.

Louis Pasteur, famed French chemist. He provided a scientific basis for part of traditional biotechnology when he showed, starting in the 1850s, that fermentation processes such as those used for millennia to produce wine, beer, buttermilk, and cheese depended on living microorganisms.

Colin Pitchfork, a 27-year-old baker in a village in Leicestershire, England, the first person to be convicted of a crime primarily on the basis of DNA "fingerprinting." When police asked for an unprecedented mass "blooding" (blood sample donation for DNA testing) of young men in several villages in an attempt to solve the rape and murder of two teenaged girls, Pitchfork tried to evade the process by having a coworker give blood in his place. He was caught, however, and his sample proved to match the semen on one of the victims. He was found guilty of the crimes in 1987.

Robert Pollack, a geneticist. He was perhaps the first to express concern about the safety of recombinant DNA experiments. Then working at Cold Spring Harbor Laboratory in New York, Pollack learned in 1971 that Paul Berg of Stanford University was planning to insert a gene from a cancer-causing virus into a type of bacterium that could infect human beings. Pollack persuaded Berg that this might be dangerous, and Berg agreed to transfer the gene between viruses but not go on to infect the bacterium.

Ingo Potrykus, scientist at the Swiss Federal Institute of Technology. He and Peter Beyer led the team that created "golden rice" in 1999. This rice contains daffodil and bacterial genes that allow it to make beta-carotene, a building block of vitamin A. Biotechnology supporters claim that eating it can prevent vitamin A deficiency, which affects 100 to 140 million children a year and makes about 500,000 of them go blind.

David Quist, researcher at the University of California, Berkeley. His paper, coauthored with Ignacio Chapela, in the November 29, 2001, issue of *Nature* caused widespread controversy when it claimed that engineered genes had appeared in native corn crops in Mexico. On April 11, 2002, *Nature* took the unprecedented step of apologizing for publishing the paper, saying that evidence did not justify the authors' conclusions. Quist, Chapela, and some other scientists stand by their report, however, and later reports by other scientists seem to confirm it.

Jeremy Rifkin, founder and president of the Foundation on Economic Trends. He has been known as a critic of biotechnology and genetic engineering since the mid-1970s. In addition to filing lawsuits and organizing protest demonstrations, Rifkin has written numerous books on the impact of scientific and technological change on society and the environment. One recent book is *The Biotech Century* (1998).

Barry Scheck, a New York defense attorney. With his partner, Peter Neufeld, he has become famous for acting in criminal cases involving identification by DNA testing and has appeared in a number of trials, including the O. J. Simpson murder trial in 1995. In 1991 they founded the Innocence Project, which has used DNA testing to show that certain convicted criminals were actually not guilty.

Matthias Schleiden, a German biologist, coauthor of the cell theory with Theodor Schwann. In 1839, the two proposed that the microscopic membrane-bound bodies called cells were the units of which all living things were made.

Theodor Schwann, German biologist, coauthor of the cell theory in 1839 with Matthias Schleiden.

Richard Seed, a Chicago physicist. He announced in January 1998 that he was planning to help infertile couples by creating human clones. Most commentators considered Seed eccentric at best and doubted that he had the ability to carry out his plan, but his announcement reignited debates on the ethics of human cloning and calls for legislation to ban such attempts.

O[renthal]. J. Simpson, a famous African-American former football star. In 1995, he went on trial for the murder of his ex-wife, Nicole Brown Simpson, and an acquaintance of hers, Ronald Goldman. Despite DNA evidence that pointed to his guilt, Simpson was acquitted, probably at least partly because his defense lawyers, who included Barry Scheck and Peter Neufeld, demonstrated that the Los Angeles police could have accidentally or deliberately contaminated evidence from the crime scene with samples of Simpson's blood.

Alfred Sturtevant, a student of Thomas Hunt Morgan, made the first chromosome maps in Morgan's laboratory at Columbia University when Sturtevant was just 19 years old. The maps showed the approximate location of genes coding for certain characteristics on fruit fly chromosomes.

Vandana Shiva, Indian physicist, author, and environmental activist. She is the director of the Research Foundation for Science, Technology and Natural Resource Policy. She strongly opposes promotion of genetically engineered crops in the developing world and large biotechnology companies' patenting of natural materials and traditional knowledge from tropical countries.

Lee J. Silver, Princeton University geneticist. Generally a supporter of human gene alteration, Silver has predicted in *Remaking Eden* (1997) and other books that, because only the wealthy will be able to afford to alter their children's genes, the altered "Gen-Rich" and unaltered "Naturals" may become so different that they will evolve into separate species.

Gregory Stock, head of the Program on Science, Technology, and Society at the University of California, Los Angeles. He supports alteration of germ-line genes with the aim of improving the human species.

Karen Sutton, one of twin sisters who applied for the job of global airline pilot with United Airlines in the 1990s (the other was Kimberly Hinton). The airline rejected her because she failed to meet its requirement for uncorrected vision, since she was extremely nearsighted without glasses, although her vision with corrective lenses was 20/20 or better. She and her sister sued the airline for violation of the Americans with Disabilities Act (ADA), but the U.S. Supreme Court rejected their suit in June 1999 because, as Sandra Day O'Connor wrote in the Court's majority opinion, the ADA requires someone to be "presently—not potentially or hypothetically—substantially limited" in a major life activity. The Court did not consider the women disabled because their disability could be corrected. This decision suggested that people with a genetic predisposition to disease but no actual illness also would not be covered by the ADA.

Robert Swanson, a venture capitalist, persuaded Herbert Boyer to join him in exploring the business possibilities of Boyer and Stanley Cohen's newly developed recombinant DNA technology. Swanson and Boyer founded Genentech (from GENetic ENgineering TECHnology) in California in 1976. When the company offered its stock to the public for the first time in 1980, the price per share jumped from $35 to $89 in the first few minutes of trading (a huge increase at the time), even though Genentech had yet to produce a product. Swanson died in December 1999.

Edward L. Tatum, a Stanford University molecular biologist, showed in 1941 that each gene in a DNA molecule coded for one enzyme (that is, a protein molecule). He and his colleague, George Beadle, won a Nobel Prize for this work in 1958.

James A. Thomson, the first person to isolate stem cells from human embryos. He did so in 1998, while at the University of Wisconsin at Madison. He worked with Geron Corporation, a California-based biotechnology company, to grow the cells in the laboratory.

Harold Varmus, a pioneer researcher on the genetics of cancer, shared a Nobel Prize with J. Michael Bishop, his co-researcher at the University of California, San Francisco, in 1989. He was head of the National Institutes of Health in Bethesda, Maryland, from 1993 to 1999. His positions on biotechnology and genetic engineering issues include a defense of research on embryonic stem cells, which uses cloned human embryos that are not allowed to develop.

Craig Venter, molecular biologist. While at the National Institutes of Health (NIH) in the early 1990s, he developed important new methods of identifying and sequencing genes. He left NIH in 1992 and has subsequently founded several biotechnology companies, including the Institute for Genomic Research (TIGR) and Celera Genomics. Celera "raced" the government-funded Human Genome Project to be the first to sequence the entire human genome; the competition is said to have ended in a tie.

Joseph Vlakovsky, a 25-year-old U.S. Marine corporal in 1995, joined Lance Corporal John C. Mayfield III in refusing to give samples of DNA for testing and archiving, as had been required of all United States military personnel since 1992. They said that keeping their DNA in a database violated their right to privacy. A court-martial in April 1996 convicted the two of refusing to obey a direct order.

Florence Wambugu, Kenyan molecular biologist. She speaks frequently about the power of genetically engineered crops to help developing-world farmers and has helped to create a genetically engineered yam able to resist a virus that destroys much of the yam crop in Africa.

James Dewey Watson, a young American studying at Cambridge University, worked with British molecular biologist Francis Crick to discover the structure of DNA in 1953. For this groundbreaking work, which ultimately showed how inherited information was coded and passed on, Watson, Crick, and Maurice Wilkins received a Nobel Prize in 1962. Watson went on to direct Cold Spring Harbor Laboratory in New York and was also the first head of the Human Genome Project. He resigned in 1992.

Ian Wilmut, a scientist at the Roslin Institute in Scotland. He startled the world in February 1997 by announcing that he and his coworkers had cloned a sheep (which they named Dolly) from a mature body cell, something many scientists had thought impossible. Wilmut was looking for a technique to create herds of identical, genetically engineered animals that could produce drugs and other substances, but his discovery spurred debates on human cloning as well. Wilmut himself has said that humans should not be cloned.

CHAPTER 5

GLOSSARY

Biotechnology, genetic engineering, and genetics are complex fields with highly technical vocabularies. This chapter presents some of the terms that the general reader is most likely to encounter while researching these fields and their ethical, legal, and social implications. Several web sites also offer online glossaries (see Chapter 6, "How to Research Biotechnology and Genetic Engineering").

ADA (adenosine deaminase) deficiency A rare inherited disorder in which lack of a certain enzyme in white blood cells causes an essentially complete failure of the immune system. It was the first illness to be treated successfully by gene therapy (in 1990).

adenine One of the four bases in DNA and RNA. It pairs with thymine in DNA and uracil in RNA.

adult stem cells Stem cells taken from an adult, fetus, or newborn (including cells from the umbilical cord and placenta). Adult stem cells probably cannot differentiate into as many different kinds of tissue as embryonic stem cells, but their use is less controversial because they can be harvested without killing living things. *See also* **stem cells.**

adverse selection The tendency of people who know they are likely to need life or health insurance, while insurers do not know of this risk, to take out large amounts of insurance, thus throwing off the insurance industry's statistical methods and raising premiums for everyone. Allowing policyholders but not insurance companies to know the results of genetic tests could increase adverse selection.

Americans with Disabilities Act (ADA) Passed in 1990, this law protects the disabled from discrimination. It covers healthy people who are perceived as disabled, which may include those with inherited defects revealed by genetic tests.

amino acids The small molecules of which proteins are made. There are 20 different kinds.

Animal and Plant Health Inspection Service (APHIS) The part of the U.S. Department of Agriculture responsible for regulating genetically altered agricultural plants and animals.

Asilomar conference A conference held at a California retreat center in February 1975, in which 140 geneticists and molecular biologists worked out safety standards for recombinant DNA experiments.

***Bacillus thuringensis* (Bt)** A bacterium that produces a toxin that kills a variety of pest insects but is harmless to most nonpest insects and other living things, including humans. Sprays containing the bacteria have been used as short-lived, "organic" insecticides. The gene that produces the Bt toxin has been inserted into several kinds of crop plants.

bases Small molecules that usually exist in pairs in DNA and RNA. In DNA, the bases are adenine, thymine, guanine, and cytosine; in RNA, uracil replaces thymine. The sequence of bases in a nucleic acid molecule contains the genetic code.

biochip *See* **DNA chip.**

bioinformatics The application of computer and statistical techniques, especially database management, to the organization of biological information, such as DNA or protein sequences.

bioprospecting Searching an environment for living things or parts of living things (including genetic material) that may have commercial use. Critics call this process "biopiracy."

bioremediation A branch of biotechnology that uses living things, usually microorganisms, to repair environmental damage, for instance by breaking down oil or other toxic chemicals.

biotechnology Using or altering living things in processes that benefit humankind; now frequently applied to commercial processes that use organisms with altered genes.

bovine growth hormone (bovine somatotropin, BGH, BST) A cattle hormone that can be made by bacteria containing recombinant DNA (the recombinant hormone is known as rBGH). The hormone is given to dairy cattle to increase milk production. Its use is controversial.

blastocyst A stage of early embryonic development (about five days after fertilization) in which the embryo consists of a ball of about 200 cells. This is the stage at which embryonic stem cells are harvested, destroying the embryo.

BRCA1, BRCA2 Genes found in the early 1990s to be inherited in some families in which breast and ovarian cancer are unusually common and occur at an early age. Five to 10 percent of women with breast cancer have a mutated form of one of these genes, which can be detected by a test.

Bt *See* ***Bacillus thuringensis.***

Glossary

carrier An organism that has inherited a gene related to a disease and therefore can pass it on to offspring but does not suffer from the disease.

Cartagena Biosafety Protocol An international treaty (actually worked out in Montreal in January 2000 after a meeting in Cartagena, Colombia, in 1999 had deadlocked) that regulates international movement of genetically modified crops in ways that minimize their threat to biodiversity; it is an addendum to the United Nations Convention on Biological Diversity (1992). The treaty went into effect on September 11, 2003.

cell The basic unit of which all organisms are composed, made up of a microscopic piece of living material surrounded by a membrane. It is the simplest living system that can exist independently.

cell line A group of cells altered so that they will multiply indefinitely in culture in a laboratory. All cells in a cell line are descended from a single cell and usually are genetically identical.

chromosome One of a group of threadlike bodies containing the DNA that carries a cell's basic genetic information (they also contain protein). In cells with nuclei, they are found in pairs inside the nucleus. Chromosomes ("colored bodies") reproduce themselves during cell division.

clone A gene, cell, or organism that is the exact genetic duplicate of another gene, cell, or organism; both are produced asexually from the same ancestor. Genes or other stretches of DNA may be cloned in bacteria for study or for biotechnology processes. Plant and (to a more limited extent) animal clones are used in biotechnology and research. The ethics of cloning humans has been hotly debated.

cold cases Crimes committed years or even decades previously, for which there is insufficient evidence to justify an arrest (and often no leads at all). Many of these cases are now being solved by comparing DNA in samples taken from the crime scenes to DNA profiles of convicted criminals stored in local, state, or national databases.

Combined DNA Index System (CODIS) Software developed by the FBI to coordinate information from state and national databases of DNA profiles from criminals and crime scenes.

Cry9C A protein produced by a gene from *Bacillus thuringensis* (Bt) inserted into a type of corn called StarLink, which was not approved for human food because Cry9C, unlike other Bt proteins, does not break down quickly in the human digestive tract. This slow breakdown increases the chances that people might develop allergic reactions to the protein.

cytoplasm The living material in the main body of a cell (outside the nucleus in nucleated cells).

cytosine One of the four bases in DNA and RNA. It always pairs with guanine.

differentiation The process of maturation in which a cell takes on characteristics associated with a particular type of tissue, such as nerve tissue or muscle tissue. Differentiation is usually, but not always, irreversible.

Directive 98/44 A European Union (EU) directive on the legal protection of biotechnological inventions, passed by the EU Parliament and Council on July 6, 1998, which excludes from patenting (on ethical grounds) processes for cloning or modifying the genetic identity of human beings, use of human embryos for industrial purposes, and processes for modifying the genetic identity of animals that may cause them suffering unless such processes are likely to produce substantial medical benefits for humans. The directive also states that individual patents may be denied if they are deemed to violate moral or ethical standards. It permits patenting of gene sequences, but only for particular industrial applications.

DNA (deoxyribonucleic acid) The chemical in which inherited information is encoded, except in some viruses. Each DNA molecule consists of two phosphate-sugar backbones twined around each other in the shape of a double helix. Pairs of bases (adenine, thymine, guanine, and cytosine) joined by weak hydrogen bonds stretch between the backbones like rungs on a twisted ladder.

DNA chip (gene chip, biochip) A small piece of glass or other material embossed with fragments of known genes (often hundreds of them on a single chip). When a drop of sample is placed on the chip, the active genes in the sample are bound by matching fragments on the chip. The matches are revealed when the chip is placed in an electronic detector, thereby showing which genes in the sample are active. Affymetrix invented this device, which it called the GeneChip, in 1996.

DNA "fingerprinting" (profiling, identification testing) A technique, invented by Alec Jeffreys of the University of Leicester in Great Britain in the early 1980s, that uses stretches of DNA that differ considerably from person to person as a means of determining whether a sample of DNA came from a particular person. It is most often used to determine family relationships or to find out whether a sample of DNA from a crime scene could have come from a certain suspect.

DNA Identification Act Passed in 1994, this law gives the FBI responsibility for overseeing training, funding, and proficiency testing of all laboratories doing forensic DNA profiling. It also authorizes establishment of a national database of DNA profiles.

Dolly A sheep (named after Dolly Parton) cloned from an udder cell of an adult ewe by Ian Wilmut and his colleagues at the Roslin Institute in Scotland. She is the first mammal to be cloned from a mature adult cell. Her existence was announced in an article in *Nature* on February 27,

1997. Dolly was euthanized in February 2003 because she had developed a virus-induced lung tumor.

dominant gene A gene that is expressed even if only one copy of it is inherited. *Compare with* **recessive gene.**

double helix The shape of a DNA molecule, in which two parallel strands twist or coil like a corkscrew.

embryonic stem cells Stem cells taken from an embryo, usually at the blastocyst stage. Embryonic stem cells have greater capacity to differentiate into different types of tissue than adult stem cells; they probably can form any tissue in the body. Many researchers therefore would rather work with embryonic than adult stem cells, but their use is more controversial because embryos must be destroyed in order to harvest them. *See also* **stem cells.**

enzyme A protein that catalyzes a chemical reaction, speeding the rate at which it occurs.

Escherichia coli (E. coli) A common type of bacterium that usually lives harmlessly in the human intestine (though some strains of it can cause serious illness). It grows easily in the laboratory and was frequently used in early recombinant DNA and other genetics experiments.

eugenics A pseudoscience founded in the late 19th century by Francis Galton (he coined the word, from Greek words meaning "well born," in 1883). It holds that complex behaviors and characteristics such as intelligence are inherited and that the human race can be improved by encouraging people with desirable traits to reproduce and discouraging or forcibly preventing those with undesirable traits from doing so.

eukaryote An organism made up of one or more cells with nuclei. All living things except viruses, bacteria, and blue-green algae are eukaryotes. *Compare with* **prokaryote.**

exon The part of a gene that carries the code for making a protein. *Compare with* **intron.**

Federal Food, Drug, and Cosmetic Act (FFDCA) First passed in 1938, this act gave the FDA and, later, the EPA the right to set tolerances for certain substances on and in food and feed. Herbicide residues on genetically engineered herbicide-tolerant food crops and pesticides (such as Bt toxin) produced by genetically engineered food crops are regulated under the FFDCA.

Federal Insecticide, Fungicide, and Rodenticide Act (FIFRA) A law, passed in 1947, that governs the testing and use of pesticides. After the Environmental Protection Agency (EPA) was established in 1970, it took over the administration of regulations required by FIFRA. FIFRA has been amended to include genetically engineered crops that produce Bt *(Bacillus thuringensis)* toxin in its definition of pesticides.

fermentation A group of processes in which microorganisms such as bacteria or yeasts cause chemical changes that include the production of gas. Fermentation is part of many traditional biotechnology processes, such as those used to make alcoholic beverages. It was thought to be strictly a chemical reaction until the mid-19th century, when Louis Pasteur proved that it required living microorganisms.

Flavr Savr A type of tomato that a company called Calgene genetically altered to delay its breakdown after ripening. In 1994, Flavr Savr became the first genetically modified food to go on sale.

"Frankenfoods" Disparaging term used for foods containing genetically modified material, especially in Europe. It refers to the monster created by a scientist in Mary Shelley's novel *Frankenstein* (1818).

Frye rule A rule (first stated in a 1923 case, *Frye v. United States*) that judges often use to decide whether evidence based on new scientific techniques will be admitted in court. The rule recommends accepting such evidence if the technique on which it is based has "gained general acceptance" among scientists in its field.

gene The basic unit of inherited information, consisting of a sequence of bases in a DNA molecule that carries the code for production of a specific protein or RNA molecule.

gene chip *See* **DNA chip.**

gene flow Transfer of genes from one population to another, such as the transfer of genes from bioengineered crops to wild relatives through processes such as the spread of pollen.

Genentech The first biotechnology company to be based on recombinant DNA technology. Herbert Boyer and Robert Swanson founded it in 1976.

gene splicing (recombinant DNA technology) Common term for insertion of one or more genes from one species into the genome of another.

gene therapy Treatment of a disease by altering genes. Gene therapy was first successfully used on a human being in September 1990.

genetic code The sequence of bases in a DNA or RNA molecule that determines the sequence of amino acids in a protein. Each "letter" of the code consists of a group of three bases.

genetic determinism The belief that most physical and mental characteristics, including complex behaviors, are determined primarily or exclusively by genetics (heredity).

genetic engineering Direct manipulation of genetic information or transfer of genes from one type of organism to another to produce new biological structures or functions useful to human beings. Genetic engineering includes, but is not limited to, recombination of DNA.

genetics The study of the patterns and mechanisms by which traits are inherited (passed from parents to offspring).

genetic screening Testing of a population to identify people at risk for suffering from a genetic disease or passing such a disease to their children.

genome An organism's complete collection of genetic information.

genomics The science of identifying genes and their functions, including building maps and databases of genes. Genomics studies large numbers of genes, or whole genomes, at once.

genotype The nature of an organism or group as determined by its genes.

germ-line genes Genes that are contained in the sex cells (the cells that will become sperm and eggs) and therefore can be passed on to offspring.

glyphosate A common type of herbicide (weed killer). Monsanto Corporation markets it under the name Roundup and has genetically engineered certain crops to be resistant to it. *See also* **"Roundup Ready" crops.**

GM (genetically modified) foods Foods that contain, or may contain, transgenic organisms.

golden rice Rice engineered in 1999 to contain daffodil and bacterial genes that allow it to make beta-carotene, a building block of vitamin A. Biotechnology supporters have hailed it as a possible cure or preventive for vitamin A deficiency, a widespread health problem for children in developing countries.

guanine One of the four bases in DNA and RNA. It always pairs with cytosine.

Harvard Oncomouse A type of mouse genetically engineered to be unusually susceptible to cancer. Intended for use in medical research and testing of possible carcinogens, it was patented by Harvard in 1988. It was the first genetically altered animal to be patented.

Health Insurance Portability and Accountability Act (HIPAA) Passed in 1996, this act includes a provision forbidding health insurers that issue group plans to employers from denying insurance on the basis of preexisting genetic conditions. It does not apply to individual plans or to employers who ensure their own workers.

hemoglobin The iron-containing pigment in red blood cells that gives the cells (and the blood) their color and carries oxygen throughout the body. Several inherited blood diseases, including sickle-cell disease, produce abnormal forms of hemoglobin.

Human Genome Project An international project, launched in 1990, that aims to sequence all the genes in the human genome. It announced a draft version of the human genome in June 2000 and a complete sequence of the genome in April 2003.

Huntington's disease An inherited form of incurable brain degeneration caused by a single dominant gene. Affected people usually show no symptoms until middle age. A person who inherits even one copy of the gene will develop the disease, and the child of someone with the disease has a

50 percent chance of developing it. The gene that causes the disease was identified in 1993 and can be detected by a test.

hybrid An offspring of parents that differ significantly in genetic nature.

Innocence Project A project established in 1991 by New York City lawyers Barry Scheck and Peter Neufeld to obtain DNA testing for convicted felons who declared their innocence. As of mid-2003, the project has freed 131 people, some of whom had been sentenced to death.

intron A DNA base sequence that interrupts the protein-coding portion of a gene (exon). The function of introns is currently unknown, so they are sometimes called "junk DNA."

in vitro fertilization Combination of an egg and a sperm in a laboratory to create an embryo that is then implanted in a uterus; often called test-tube fertilization. ("In vitro" is Latin for "in glass.")

"junk DNA" Popular name for introns. *See also* **intron.**

ligase A type of enzyme that permanently links complementary single strands of DNA.

locus (pl. *loci*) The place on a chromosome occupied by a gene or other specified sequence of DNA used as a marker.

maize Alternate name for the crop commonly called corn in the United States.

mitochondrial DNA DNA contained within mitochondria, organelles that help a cell use energy. Unlike nuclear DNA, mitochondrial DNA is inherited only through the female line. It has been used to trace family relationships and track missing persons. It can also be used for identification when nuclear DNA is not available, for instance in skeletal remains. Some diseases can be carried in mitochondrial DNA, and special in vitro fertilization techniques have been developed to avoid transmission of these illnesses. Such techniques may inadvertently change germ-line genes.

molecular biology A branch of biology dealing with the molecular basis of biological processes such as protein synthesis and transmission of inherited characteristics.

monoclonal antibodies Biotechnology drugs designed to home in on particular kinds of cells, such as cancer cells, and kill them without harming other cells. Monoclonal antibodies were hailed as possible "miracle drugs" in the early 1980s, but their performance in human tests remained disappointing for about 20 years. Two types began to be sold as treatments for certain kinds of cancer in 1999, and more were expected to follow.

moratorium A temporary halt to an activity.

mutation An inheritable change in a DNA sequence.

National DNA Index System (NDIS) A national database of DNA identity profiles of convicted criminals, established by the FBI in October 1998.

National Institutes of Health (NIH) A group of large, prestigious medical research institutions funded by the U.S. government and located in Bethesda, Maryland.

natural selection The mechanism of evolution described by Charles Darwin in *On the Origin of Species* (1859) and independently by Alfred Russel Wallace. It states that the members of a species with inherited characteristics that make them most suited to survive in a particular environment will be most likely to survive and bear young. Over generations, therefore, a change in the environment can cause a change in the predominant characteristics of a species.

neem A tree native to India, products of which have been used for millennia to fight bacteria, fungi, and pest insects. Multinational giant W. R. Grace and the USDA obtained a European patent for using neem oil against fungi on plants in 1994, which activists considered a "biopiracy" theft of traditional people's knowledge. The European Patent Office revoked the patent in 2000.

nuclear transfer A technique for cloning in which the nucleus of a cell (embryonic or, more recently, mature) is transferred into an unfertilized egg from which the nucleus has been removed. The two are then fused by electricity. The resulting single cell develops into a clone of the organism that provided the nucleus.

nucleic acid A large molecule composed of nucleotides. The most common nucleic acids are DNA and RNA.

nucleotide A subunit of DNA or RNA composed of a base and an attached "backbone" of phosphate and sugar. Each DNA or RNA molecule contains thousands of nucleotides linked together.

nucleus The membrane-bound central body in eukaryotic cells that contains the cell's main genetic material.

PCR *See* **polymerase chain reaction.**

pharming Use of genetically altered farm animals or plants to produce human body chemicals and drugs.

Plant Patent Act Passed in 1930, this act stops short of allowing conventional patents on plants but allows plant breeders to forbid others to clone (reproduce asexually) hybrid varieties that they have developed.

Plant Pest Act Passed in 1987, this act gives the U.S. Department of Agriculture's Animal and Plant Health Inspection Service (APHIS) the right to regulate any organisms, including those produced by genetic engineering, that are or might be plant pests.

Plant Variety Protection Act This act, passed in 1970, extends the Plant Patent Act to forbid sexual reproduction of plant varieties without their developers' permission.

plasmid A circular DNA molecule, used by some bacteria to transfer genes from one microorganism to another. It is separate from the main bacterial genome and can reproduce on its own. Herbert Boyer and Stanley Cohen used plasmids in some of the first recombinant DNA experiments.

polymerase chain reaction (PCR) A method for rapidly multiplying copies of a DNA sequence, developed in the early 1980s by Kary Mullis. It can be used to increase tiny samples of DNA to a size usable in DNA identification testing or gene sequencing.

precautionary principle The idea that a new technology should be assumed to be, and regulated as, potentially dangerous until it has been proved safe, even if the dangers have not been proved. The Cartagena Biosafety Protocol and numerous national and international laws and treaties apply this principle to genetically modified foods.

preimplantation genetic diagnosis (PGD) A technique developed in the early 1990s, in which a group of embryos are created from a couple's eggs and sperm in a fertility clinic and then analyzed genetically. Only embryos that meet specified criteria, such as absence of a disease-causing gene or presence of genes producing tissue compatibility with a sibling, are implanted and allowed to develop into babies.

prokaryote A cell or organism that lacks a separate nucleus, such as a bacterium. *Compare with* **eukaryote.**

protein One of a large family of substances that are composed of amino acids arranged in a certain order. Proteins carry out most functions in cells, including acting as enzymes, structural components, and signaling molecules. They are assembled according to instructions in the genetic code contained in DNA and RNA. Instructions for each type of protein are carried on a separate gene.

rBGH *See* **bovine growth hormone.**

proteomics The study of the structure and function of the proteins that are produced by genes.

reach-through rights Rights demanded by some holders of biotechnology patents as part of licensing agreements. If a company has reach-through rights to a technology, it can demand royalties on all products developed with that technology, even if the development work is done by others.

recessive gene A gene that does not produce a detectable characteristic unless copies of that gene have been inherited from both parents. *Compare with* **dominant gene.**

recombinant DNA DNA from different types of organisms that is combined directly in a laboratory.

Recombinant DNA Advisory Committee (RAC) A group of experts formed by the National Institutes of Health in October 1974 to judge the safety of recombinant DNA experiments. It still exists, but it now reviews only experiments that differ from previous ones in a substantial way or involve treatment of humans.

reproductive cloning Cloning of human embryos or cells for the purpose of producing a child. *Compare with* **research cloning.**

research cloning Cloning for the purpose of producing embryos from which embryonic stem cells can be harvested; the embryos are never allowed to develop into babies. It is also called therapeutic cloning. *Compare with* **reproductive cloning.**

restriction enzyme (restriction endonuclease) An enzyme produced by some bacteria to break up the DNA of invading viruses. A restriction enzyme cuts a DNA molecule at any spot where a particular sequence of bases occurs, leaving a piece of single-stranded DNA at each end. There are many types of restriction enzymes, each of which cuts at a different sequence. Restriction enzymes have been used in recombinant DNA technology since its start.

RNA (ribonucleic acid) A single-stranded nucleic acid with a molecular structure similar to that of DNA but with uracil substituting for thymine among the bases. Unlike DNA, RNA is found in the cytoplasm of the cell as well as the nucleus. It plays a vital role in turning the DNA code into protein molecules and in other chemical activities of the cell. It exists in several forms, including messenger RNA and transfer RNA.

"Roundup Ready" crops Crops genetically engineered by Monsanto Corporation to be resistant to glyphosate, a type of herbicide that the company markets under the name Roundup.

short tandem repeats (STRs) Areas of DNA containing many short, identical sequences in a row. The number of repeats varies considerably from person to person. A test using STRs at 13 loci has become the standard form of DNA profiling in the United States.

sickle-cell disease (sickle-cell anemia) An inherited blood disease, fairly common among people of African descent, that is caused by abnormal hemoglobin produced by a defective gene. The mutant hemoglobin deforms the round cells to a sickle shape and causes them to block tiny blood vessels, starving the body of oxygen and causing pain, illness, and sometimes an early death. Because the gene that causes the disease is recessive, people develop it only if they inherit the mutant gene from both parents. A person with only one mutant gene is called a sickle-cell carrier or a possessor of sickle-cell trait.

sickle-cell trait A trait possessed by people, usually of African descent, who inherit one normal and one mutant copy of the gene that, when a person inherits two copies, causes sickle-cell anemia. People with sickle-cell trait are also called sickle-cell carriers because they can pass the defective gene to their children. They themselves, however, are perfectly healthy. Indeed, they seem to have unusual resistance to malaria, a serious blood disease caused by a parasite that is widespread in Africa. People with sickle-cell trait have sometimes been discriminated against in employment or insurance because of the mistaken impression that they are or will become ill.

somatic cell Any cell in the body except the sex cells (sperm and eggs) and their precursors.

somatic cell nuclear transfer A cloning technique in which the cloned cell is a mature body cell. This technique was used to create Dolly the sheep. *See also* **nuclear transfer.**

StarLink A type of genetically engineered corn approved for animal feed but not for human food because of the possibility that it might cause allergic reactions. Small amounts of it were found to have tainted corn-containing foods in fall 2000, causing a wide outcry and a recall of 430 million bushels of corn and 300 types of corn products.

stem cell A cell that has not yet differentiated and has the potential to produce cells of a wide range of types, in some cases any type. The study of stem cells may eventually produce tissues for transplantation or treatment of degenerative diseases. Stem cells from early embryos (embryonic stem cells) are thought to have the ability to develop into more different types of tissues than stem cells taken from adults, newborns, or fetuses (adult stem cells). Many scientists therefore prefer them for research, but their use is more controversial because embryos must be destroyed in order to harvest them.

"sticky ends" Pieces of single-stranded DNA at the ends of a double stranded segment left after cutting by restriction enzymes or certain other treatments. They attract and bind to other pieces of single-stranded DNA with a complementary sequence. This fact was used in the creation of recombinant DNA in the early 1970s.

structural gene A gene that contains coded instructions for making a particular protein or RNA molecule. *Compare with* **regulatory gene.**

SV40 (Simian Virus 40) A monkey virus that can cause cancer in mice. Paul Berg used it in the first recombinant DNA experiments.

technology transfer The process of converting the results of basic scientific research into commercially useful products.

therapeutic cloning An alternate name for research cloning, stressing the possible use of stem cells obtained from cloned embryos as a medical treatment. *See also* **research cloning.**

thymine One of the four bases in DNA. It always pairs with adenine. In RNA it is replaced by uracil.

Toxic Substances Control Act (TSCA) An act passed in 1976 that gives the Environmental Protection Agency (EPA) the right to regulate release and sale of "new" chemicals. TSCA was amended in 1986 to classify genetically engineered organisms expressing new traits or containing genes from two different genera as equivalent to new chemicals and thus susceptible to regulation under the act.

Trade Related Intellectual Property Rights Agreement (TRIPS) Agreement of the World Trade Organization (WTO), which went into effect in 1995. This agreement requires WTO member countries to "harmonize" their patent laws with those of the United States and other industrialized countries, supposedly in order to facilitate the transfer of technology to those countries. Critics say its real purpose is to protect the rights and profits of multinational corporations.

transgenic organism A living thing that has been changed by adding genes from another species, removing genes, or changing the activity (expression) of existing genes.

uracil One of the four bases in RNA. It is the equivalent of thymine in DNA and pairs with adenine.

vector A tool for carrying genes from one organism into the genome of another. Plasmids and certain viruses are examples of vectors.

PART II

GUIDE TO FURTHER RESEARCH

CHAPTER 6

How to Research Biotechnology and Genetic Engineering

Although students, teachers, journalists, and other investigators may ultimately have different objectives in doing research on biotechnology and genetic engineering, all are likely to begin with the same basic steps. The following approach should be suitable for most purposes:

- Gain a general orientation by reading the first part of this book. Chapter 1 can be read as a narrative, while Chapters 2–5 are best skimmed to get an idea of what is covered. They can then be used as a reference source for helping make sense of the events and issues encountered in subsequent reading.
- Skim some of the general books listed in the first section of the bibliography (Chapter 7). Neutral overviews and books that provide pro and con essays on various issues in the field are particularly recommended.
- Browse the many web sites provided by organizations involved in various aspects of biotechnology and genetics issues (see Chapter 8). Whether they favor, oppose, or are neutral toward the field, their pages are often rich in news, articles, and links to other organizations as well as describing particular cases and discussing the pros and cons of various practices involving biotechnology.
- Use the relevant sections of Chapter 7 to find more books, articles, and online publications on particular topics of interest.
- Find more (and more recent) materials by using the bibliographic tools such as the library catalogs and periodical indexes discussed later.
- To keep up with current events and breaking news, check back periodically with media and organization web sites and periodically search the catalogs and indexes for recent material.

The rest of this chapter is organized according to the various types of resources and tools. The three major categories are online resources, bibliographic resources, and legal research, including both laws and court cases.

ONLINE RESOURCES

The tremendous growth in the resources and services available through the Internet (and particularly the World Wide Web) is providing powerful new tools for researchers. Mastery of a few basic online techniques enables today's researcher to accomplish in a few minutes what used to require hours in the library poring through card catalogs, bound indexes, and printed or microfilmed periodicals.

Not everything is to be found on the Internet, of course. While a few books are available in electronic versions, most must still be obtained as printed text. Many periodical articles, particularly those more than 10 years old, must still be obtained in hard-copy form from libraries. Nevertheless, the Internet has now reached critical mass in the scope, variety, and quality of material available. Thus, it makes sense to make the Internet the starting point for most research projects. This is particularly true regarding recent events in biotechnology and genetic engineering. Web/Internet links can lead the researcher not only to companies, professional organizations, and journals in the field but even to complex databases of DNA sequences and other highly technical material.

For someone not used to it, searching the Internet and the World Wide Web can feel like spending hours trapped inside a pinball machine. The shortest distance between a researcher and what he or she wants to know is seldom a straight line, at least not a single straight line. These things are called nets and webs for good reason: Everything is connected by links, and often a researcher must travel through a number of these to find the desired information.

Web searching is best approached with a combination of patience, alertness, and, preferably, humor. A given search often will not reveal the desired information but will unearth at least three things, or groups of things, that are even more interesting. The information sought on the initial search, meanwhile, will be uncovered by chance at a later time when the researcher is looking for something else entirely. The sooner one accepts this, the sooner searching is likely to become rewarding rather than painful. In addition to specific files related to particular areas of research, it is a good idea for researchers to save promising URLs (universal resource locators or web addresses) in their web browser's Favorites list.

It is easy to feel lost on the Web, but it is also easy to find one's way back. During any given search, the Back button on the browser is the Ariadne's thread that will guide the researcher back through the labyrinth to the beginning of the adventure on the browser's home page, passing en route through all the sites visited (so that one can stop for another look or, if desired, jump off to somewhere else). Alternatively, Home (the house icon) takes users directly back to their home page, like Dorothy clicking her red shoes. The History button provides a list of all the sites visited on recent previous sessions.

Finally, a word of caution about the Internet. It is important to critically evaluate all online materials. Many sites have been established by well-known, reputable organizations or individuals. Others may come from unknown individuals or groups. Their material may be equally valuable, but it should be checked against reliable sources. Different groups' biases may be obvious or subtle. Gradually, each researcher will develop a feel for the quality of information sources as well as a trusty tool-kit of techniques for dealing with them.

TOOLS FOR ORGANIZING RESEARCH

Several techniques and tools can help the researcher keep materials organized and accessible:

Use the web browser's "Favorites" or "Bookmarks" menu to create a folder for each major research topic (and optionally, subfolders). For example, folders used in researching this book included: organizations, laws, cases, current news, reference materials, and bibliographical sources.

Use favorites or bookmark links rather than downloading a copy of the actual web page or web site, which can take up a large amount of both time and disk space. Exception: if the site has material that will definitely be needed in the future, download it to guard against its disappearance from the Web.

Use a simple database program (such as Microsoft Works) or, perhaps better, a free-form note-taking program. This makes it easy to take notes (or paste text from web sites) and organize them for later retrieval.

WEB INDEXES

A web index is a site that offers what amounts to a structured, hierarchical outline of subject areas. This organization enables the researcher to zero in on a particular aspect of a subject and find links to web sites for further exploration. A staff of researchers constantly adds to and updates the links.

The best known (and largest) web index is Yahoo! (http://www.yahoo.com). Its home page gives the top-level list of topics. Five of these are of particular use for researching biotechnology and genetic engineering:

Science: Numerous topics under this major heading are relevant. In addition to general subjects such as journals and news/media, specific secondary subject headings likely to be helpful are Agriculture; Animals, Insects, and Pets; Biology; Ecology; Forensics; Life Sciences; Medicine; and Science and Technology Policy. Selecting the secondary heading Biology produces a further host of useful subheads, including Ask an Expert, Biochemistry, Biodiversity, Biological Safety, Biomedical Ethics, Biotechnology, Cell Biology, Companies, Genetics, Institutes, Journals, Molecular Biology, Organizations, and Pharmacology.

Government: Provides listings including documents, ethics, law, and research labs. Categories under the Law subhead that are likely to be useful include Cases, Disability, Employment Law, Environmental, Health, Indigenous Peoples, Intellectual Property, Legal Research, and Technology.

Health produces subheads dealing with medicine and with diseases and conditions, which includes genetic disorders.

Headings under Business and Economy that may be useful include Ethics and Responsibility, Global Economy, Intellectual Property, and Law.

Society and Culture may have biotechnology information under the headings of Disabilities, Environment and Nature, and Issues and Causes as well as more general headings such as Cultural Policy.

In addition to following Yahoo!'s outline-like structure, there is also a search box into which the researcher can type one or more keywords and receive a list of matching categories and sites. (The box is rather confusingly labeled "Search the Web," but it also searches Yahoo!'s directories, and the results of this search appear at the top of the page.)

Web indexes such as Yahoo! have two major advantages over undirected surfing. First, the structured hierarchy of topics makes it easy to find a particular topic or subtopic and then explore its links. Second, Yahoo! does not make an attempt to compile every possible link on the Internet (a task that is virtually impossible, given the size of the Web). Rather, sites are evaluated for usefulness and quality by Yahoo!'s indexers. This means that the researcher has a better chance of finding more substantial and accurate information. The disadvantage of Web indexes is the flip side of their selectivity: The researcher is dependent on the indexer's judgment for determining what sites are worth exploring.

Two other Web indexes are LookSmart (http://www.looksmart.com) and About.com (http://www.about.com), the latter of which is run by About, a company formerly named The Mining Company. About.com features overviews or guides prepared by self-declared experts.

SEARCH ENGINES

Search engines take a very different approach to finding materials on the Web. Instead of organizing topically in a "top down" fashion, search engines work their way "from the bottom up," scanning through Web documents and indexing them. There are hundreds of search engines, but some of the most widely used are the following:

AltaVista (http://www.altavista.com)
Excite (http://www.excite.com)
Go.com (http://infoseek.go.com)
Google (http://www.google.com)
Hotbot (http://www.hotbot.com)
Lycos (http://www.lycos.com)
WebCrawler (http://www.webcrawler.com)

Search engines are generally easy to use by employing the same sorts of keywords that work in library catalogs. There are a variety of Web search tutorials available online (try "Web search tutorial" in a search engine).

Here are a few basic rules for using search engines:

When looking for something specific, use the most precise term or phrase. For example, when looking for information about DNA fingerprinting, use "DNA fingerprinting" or "forensic DNA testing," not "DNA." (When using phrases as search specifications, enclose them in quotation marks.)

When looking for a more general topic, use several descriptive words (nouns are more reliable than verbs), such as privacy genetic information. (Most engines will automatically put pages that match all three terms first on the results list.)

Use "wildcards" when a desired word may have more than one ending. For example, gene* matches genetics, genetic engineering, genome, and so on.

Most search engines support Boolean (AND, OR, NOT) operators, which can be used to broaden or narrow a search.

Use AND to narrow a search. For example, agriculture AND biotechnology will match only pages that have both terms.

Use OR to broaden a search. Agriculture OR biotechnology will match any page that has *either* term.

Use NOT to exclude unwanted results. Biotechnology NOT agriculture finds articles about biotechnology other than agricultural biotechnology.

Since each search engine indexes somewhat differently and offers somewhat different ways of searching, it is a good idea to use several different search engines, especially for a general query. Several "metasearch" programs automate the process of submitting a query to multiple search engines. These include:

Dogpile (http://www.dogpile.com)
Metacrawler (http://www.metacrawler.com)
CNET Search (http://www.search.com)
Surfwax (http://www.surfwax.com)

Metasearch engines may overwhelm researchers with results (and may insufficiently prune duplicates), however, and they often do not use some of the more popular search engines, such as Google and Northern Light.

MEDIA SITES

News (wire) services, most newspapers, and many magazines have web sites that include breaking news stories and links to additional information. The following media sites have substantial listings for stories on biotechnology and genetic engineering:

- ABC News: URL: http://abcnews.go.com
- Cable News Network (CNN): URL: http://www.cnn.com
- *New York Times:* URL: http://www.nytimes.com (offers only abstracts unless readers pay)
- Reuters: URL: http://www.reuters.com
- *Time magazine:* URL: http://www.time.com/time
- *Washington Post:* URL: http://www.washingtonpost.com

Yahoo! maintains a large set of links to many newspapers that have web sites or online editions: http://dir.yahoo.com/News_and_Media/Newspapers/Web_Directories.

METASITES: EVERYTHING ABOUT BIOTECH AND THEN SOME

One basic principle of research is to take advantage of the fact that other people may have already found and organized much of the most useful information about a particular topic. For biotechnology and genetic engineering, there are several web sites that can serve as excellent starting points for research because they provide links to vast numbers of other resources.

- Access Excellence, http://www.accessexcellence.org, is aimed at biology teachers and students. It has an extensive "About Biotech" page at http://www.accessexcellence.org/AB. Subsections of this page include Issues and Ethics, Biotech Applied, Careers, Graphics Gallery, and Biotech Chronicles (history).

- Deakin University's Biological and Chemical Sciences Department has a good list of links on subjects in biology and chemistry, including FBI DNA testing, the Human Genome Project, and human genome maps, at http://www.deakin.edu.au/fac_st/bcs/goodies/biol.html.
- The Genetic Education Center of the University of Kansas Medical Center, http://www.kumc.edu/gec, has links related to the Human Genome Project, genetic conditions, educational resources and activities, glossaries, and more.
- The Institute for Genomic Research (TIGR) provides a links page at http://www.tigr.org/links. The links cover news, organizations, genome maps and sequences, and more.
- Indiana University and the University of Texas have a combined site, Biotech: Life Sciences Resources and Reference Tools, at http://biotech.icmb.utexas.edu. It includes an illustrated dictionary, an extensive list of Science Resources links, and BioMedLink, a mega-database of biomedical sites.
- Cato Research, Ltd., at http://www.cato.com/biotech, provides the World Wide Web Virtual Library's Information Directory Section for biotechnology, covering biotech, drug development, genomics, and related fields. Most sites appear to be professional/technical.
- The U.S. Food and Drug Administration (FDA) web site has a page on genetically engineered foods, http://www.cfsan.fda.gov/~lrd/biotechm.html, and one on gene therapy, http://www.fda.gov/cber/gene.htm. The FDA's Center for Veterinary Medicine, which regulates transgenic and cloned animals and animal feeds, has a site describing the regulations at http://www.fda.gov/cvm/biotechnology/bioengineered.html.
- The international Organization for Economic Cooperation and Development (OECD) has considerable material about biotechnology at http://www.oecd.org/topic/0,2686,en_2649_37437_1_1_1_1_37437,00.html.
- GeneWatch UK, http://www.genewatch.org, a British organization that is generally negative toward genetic modification, provides access to a wide variety of news stories and reports on the subject. Topics covered include genetically modified crops and foods, genetically modified animals, human genetics, laboratory use, biological weapons, and patenting. Some reports are not available online and must be ordered.

SPECIFIC INTERNET/WEB RESOURCES

In addition to the metasites, there are many web pages devoted to particular aspects of biotechnology and genetic engineering. Here is a small sampling of the most interesting and extensive ones:

- Foundations of Classical Genetics, http://www.esp.org/foundations/genetics/classical, offers original manuscripts of some of the most important papers in the history of genetics research. It also has a chronology and links to related sites.
- The University of Reading in Great Britain has an excellent site at http://www.ncbe.reading.ac.uk/NCBE/GMFOOD/menu.html, offering British government and other reports and regulations related to genetically modified foods, a very controversial subject in Europe.
- The National Human Genome Research Institute, which carries out the Human Genome Project in the United States, has a variety of resources related to the project on its site at http://www.genome.gov. Its Policy and Ethics section has material on privacy of genetic information, genetic discrimination in employment and insurance, commercialization and patenting, DNA forensics, and genetics and the law. The project's Ethical, Legal and Social Implications (ELSI) Research Program has its own site at http://www.genome.gov/10001618.
- The National Center for Biotechnology Information, http://www.ncbi.nlm.nih.gov, provides access to a variety of human genome and other gene sequence databases.
- The U.S. Department of Agriculture's Animal and Plant Health Inspection Service site, http://www.aphis.usda.gov/brs/, contains detailed information on how APHIS regulates the movement, importation, and field testing of genetically engineered plants and microorganisms through permitting and notification procedures. Other APHIS sites accessible from this site contain information on permits for other types of genetically engineered organisms or products such as transgenic arthropods, products with applications as veterinary biologics, and articles on related subjects such as plant pests, plants and plant products, and animal and animal products.
- Information Systems for Biotechnology: Agbiotech Online, http://www.isb.vt.edu/regulatory.cfm, also includes regulations from APHIS, as well as the 1997 revision of NIH guidelines for research involving recombinant DNA molecules and other information on regulatory assessment, news reports, research resources, and links.
- The National Agricultural Library and the University of Maryland host an agricultural biotechnology site at http://agnic.umd.edu. This site includes biotechnology resources as they relate to domestic animals, plants, and food processing. Types of resources include links to portals, indexes, or gateways; organizations such as government agencies, international organizations, and scientific societies and associations; and mailing lists and discussion groups, databases, reports, and publications. Commercial sites are included if they offer free Internet resources and services.

- The Union of Concerned Scientists, http://www.ucsusa.org, has an extensive links page on agricultural biotechnology, mostly listing groups opposed to the technology, at http://www.ucsusa.org/food_and_environment/biotechnology/page.cfm?pageID=804.
- Ag BioTech InfoNet, http://www.biotech-info.net, sponsored by an international coalition of environmental, scientific, and consumer groups, covers all aspects of the application of biotechnology and genetic engineering in agricultural production and food processing and marketing. Its focus is on scientific reports and findings and technical analysis, but the page also covers emerging issues of widespread interest, developments in the policy arena, and major media coverage.
- Three sites strongly supportive of agricultural biotechnology, sponsored by or linked to leading biotechnology companies and trade organizations, are Bio-Scope (http://www.bio-scope.org), AgBioWorld (http://www.agbioworld.org), and Council for Biotechnology Information (http://www.whybiotech.com).
- For a more neutral view of agricultural biotechnology, see the Pew Initiative on Food and Biotechnology, http://pewagbiotech.org, which contains numerous reports, papers, and news items. The United Nations Food and Agriculture Organization (FAO) hosts the Forum on Biotechnology in Food and Agriculture at http://www.fao.org/biotech/index.asp? lang=en.
- Information about the FBI's CODIS (Combined DNA Index System) program can be found at http://www.fbi.gov/hq/lab/codis/index1.htm. NDIS, the national forensic DNA database, is described more briefly at http://foia.fbi.gov/dna552.htm.
- The Reproductive Cloning Network, http://www.reproductivecloning.net/, and The Human Cloning Foundation, http://www.humancloning.org, contain essays explaining and supporting human reproductive cloning (cloning to produce children), as well as lists of books and other resources on the subject.
- Cells Alive, http://www.cellsalive.com, produces pictures, including some videos, of different kinds of cells, crystals and more, with explanatory text. It enables the viewer to see heart cells beating onscreen, for example. It is a good educational resource.

FINDING ORGANIZATIONS AND PEOPLE

Lists of organizations connected with biotechnology can be found on archive sites (see Specific Internet/Web Resources, p. 155) and index sites such as Yahoo!, as well as in Chapter 8 of this book. If such sites do not yield the name of a specific organization, the name can be given to a search engine. Put the name of the organization in quotation marks.

Another approach is to take a guess at the organization's likely Web address. For example, the American Civil Liberties Union (which includes genetic discrimination among its concerns) is commonly known by the acronym ACLU, so it is not a surprise that the organization's web site is at http://www.aclu.org. (Note that noncommercial organization sites normally use the .org suffix, government agencies use .gov, educational institutions have .edu, and businesses use .com.) This technique can save time, but it does not always work.

There are several ways to find a person on the Internet:

Put the person's name (in quotes) in a search engine and possibly find that person's home page on the Internet.

Contact the person's employer (such as a university for an academic, or a corporation for a technical professional). Most such organizations have web pages that include a searchable faculty or employee directory.

Try one of the people-finder services such as Yahoo! People Search (http://people.yahoo.com) or BigFoot (http://www.bigfoot.com). This may yield contact information such as an e-mail address, regular address, and/or phone number.

BIBLIOGRAPHIC RESOURCES

Bibliographic resources generally include catalogs, indexes, bibliographies, and other guides that identify the books, periodical articles, and other printed resources that deal with a particular subject. They are essential tools for the researcher.

LIBRARY CATALOGS

Most public and academic libraries have replaced their card catalogs with online catalogs, and many institutions now offer remote access to their catalog, either through dialing a phone number with terminal software or connecting via the Internet.

Access to the largest library catalog, that of the Library of Congress, is available at http://catalog.loc.gov. This page explains the different kinds of catalogs and searching techniques available.

Yahoo! offers a categorized listing of libraries at http://dir.yahoo.com/reference/libraries. Of course, one's local public library (and for students, the high school or college library) is also a good source for help in using online catalogs.

With traditional catalogs, lack of knowledge of appropriate subject headings can make it difficult to make sure the researcher finds all relevant materials. Online catalogs, however, can be searched not only by author, title, and subject but also by matching keywords in the title. Thus a title search for "biotechnology" will retrieve all books that have that word somewhere in their title. (Of course, a book about biotechnology may not have the word *biotechnology* in the title, so it is still necessary to use subject headings to get the most comprehensive results.)

Once the record for a book or other item is found, it is a good idea to see what additional subject headings and name headings have been assigned. These in turn can be used for further searching.

BOOKSTORE CATALOGS

Many people have discovered that online bookstores such as Amazon.com (http://www.amazon.com) and Barnes & Noble (http://www.barnesandnoble.com) are convenient ways to shop for books. A less-known benefit of online bookstore catalogs is that they often include publisher's information, book reviews, and readers' comments about a given title. They can thus serve as a form of annotated bibliography.

On the other hand, a visit to one's local bookstore also has its benefits. While the selection of titles available is likely to be smaller than that of an online bookstore, the ability to physically browse through books before buying them can be very useful.

PERIODICAL DATABASES

Most public libraries subscribe to database services, such as InfoTrac, which index articles from hundreds of periodicals, including some specialized ones. The database can be searched by author or by words in the title, subject headings, and sometimes words found anywhere in the article text. Depending on the database used, "hits" in the database can result in just a bibliographical description (author, title, pages, periodical name, issue date, and so on), a description plus an abstract (a paragraph summarizing the contents of the article), or the full text of the article itself.

Many libraries provide dial-in or Internet access to their periodical databases as an option in their catalog menu. However, licensing restrictions usually mean that only researchers who have a library card for that particular library can access the database (by typing in their name and card number). Check with local public or school libraries to see what databases are available.

A somewhat more time-consuming alternative is to find the web sites for magazines likely to cover a topic of interest. Some scholarly publications are putting all or most of their articles online. Popular publications tend to offer only a limited selection. Some publications of both types offer archives of several years' back issues that can be searched by author or keyword.

Nearly all newspapers now have web sites with current news and features. Generally a newspaper offers recent articles (perhaps from the last 30 days) for free online access. Earlier material can often be found in an archive section. A citation and perhaps an abstract is frequently available for free, but a fee of a few dollars may be charged for the complete article. One can sometimes buy a "pack" of articles at a discount as long as the articles are retrieved within a specified time. Of course, back issues of newspapers and magazines may also be available in hard copy, bound, or microfilm form at local libraries.

LEGAL RESEARCH

As issues related to biotechnology, genetic engineering, and human genetics continue to capture the attention of legislators and the public, a growing body of legislation and court cases is emerging. Because of the specialized terminology of the law, legal research can be more difficult to master than bibliographical or general research. Fortunately, the Internet has also come to the rescue in this area, offering a variety of ways to look up laws and court cases without having to pore through huge bound volumes in law libraries (which may not be accessible to the general public, anyway). To begin with, simply entering the name of a law, bill, or court case into a search engine will often lead the researcher directly to both text and commentary.

FINDING LAWS

When federal legislation passes, it becomes part of the U.S. Code, a massive legal compendium. Laws can be referred to either by their popular name or by a formal citation. For example, the DNA Identification Act is cited as 42 U.S.C. 14131, meaning title 42 of the U.S. Code, section 14131.

The U.S. Code can be searched online in several locations, but the easiest site to use is probably the one from Cornell Law School (a major-provider of free online legal reference material), at http://www4.law.cornell.edu/uscode. The fastest way to retrieve a law is by its title and section citation, but phrases and keywords can also be used.

Federal laws are generally implemented by a designated agency that writes detailed rules, which become part of the Code of Federal Regulations (C.F.R.). A regulatory citation looks like a U.S. Code citation and takes the form *vol. C.F.R. sec. number*, where *vol.* is the volume number and *number* is the section number. Regulations can be found at the web site for the relevant government agency, such as the Food and Drug Administration or the Environmental Protection Agency.

Many state agencies have home pages that can be accessed through the Findlaw state resources web site (http://www.findlaw.com/11stategov). This site also has links to state law codes. These links may or may not provide access to the text of specific regulations, however.

KEEPING UP WITH LEGISLATIVE DEVELOPMENTS

Pending legislation is often tracked by advocacy groups, both national and those based in particular states. See Chapter 8, "Organizations and Agencies," for contact information.

The Library of Congress Thomas web site (http://thomas.loc.gov) includes files summarizing legislation by the number of the Congress. Each two-year session of Congress has a consecutive number; for example, the 108th Congress was in session in 2003 and 2004. Legislation can be searched for by the name of its sponsor(s), the bill number, or by topical keywords. (Laws that have been passed can be looked up under their Public Law number.) For instance, selecting the 108th Congress and typing the phrase "human cloning" into the search box at the time of writing retrieved 12 bills containing that phrase. Further details retrievable by clicking on the bill number and then the link to the bill summary and status file include text, sponsors, committee action, and amendments.

A second extremely useful site is maintained by the Government Printing Office (http://www.gpoaccess.gov/index.html). This site has links to the Code of Federal Regulations, the Federal Register (which contains announcements of new federal agency regulations), the Congressional Record, the U.S. Code, congressional bills, a catalog of U.S. government publications, and other databases. It also provides links to individual agencies, grouped under government branch (legislative, executive, judicial), and to regulatory agencies, administrative decisions, core documents of U.S. democracy such as the Constitution, and hosted federal web sites.

FINDING COURT DECISIONS

Like laws, legal decisions are organized using a system of citations. The general form is: *Party 1 v. Party 2 volume court reports (year).*

Here are some examples from Chapter 2:

Bragdon v. Abbott, 97 U.S. 156 (1998)

Here the parties are Bragdon (plaintiff) and Abbott (defendant), the case is in volume 97 of the *U.S. Supreme Court Reports*, and the year the case was decided is 1998. (For the Supreme Court, the name of the court is omitted).

John Moore v. Regents of California, 51 Cal. 3d 120 (1990)

Here the parties are John Moore (plaintiff) and the Regents of the University of California (defendant), the decision is in volume 51 of the California Supreme Court records, and the case was decided in 1990.

To find a federal court decision, first ascertain the level of court involved: district (the lowest level, where trials are normally held), circuit (the main court of appeals), or the Supreme Court. The researcher can then go to a number of places on the Internet to find cases by citation and, often, the names of the parties or subject keywords. Two of the most useful sites are the following:

The Legal Information Institute (http://supct.law.cornell.edu/supct/index.html) has all Supreme Court decisions since 1990 plus 610 of the most important older decisions. It also links to other databases with early court decisions.

Washlaw Web (http://www.washlaw.edu) has a variety of courts (including state courts) and legal topics listed, making it a good jumping-off place for many sorts of legal research. However, the actual accessibility of state court opinions varies widely.

For more information on conducting legal research, see the "Legal Research FAQ" at http://www.faqs.org/faqs/law/research. After a certain point, however, the researcher who lacks formal legal training may need to consult with or rely on the efforts of professional researchers or academics in the field.

CHAPTER 7

ANNOTATED BIBLIOGRAPHY

Hundreds of books and thousands of articles and Internet documents relating to biotechnology, genetic engineering, and genetics have appeared in recent years as these fields have grown in complexity and importance. They range from extremely technical "how-to" articles and descriptions of particular advances to reviews and opinion pieces aimed at the general public. This bibliography lists a representative sample of serious nonfiction sources selected for clarity and usefulness to the general reader, recent publication (except for some items containing material of historical interest, most material dates from 2000 or later), and variety of points of view.

Listings in this bibliography are grouped by subject under the following headings:

- General Material on Biotechnology and Genetics
- Agricultural Biotechnology and Safety
- Patenting Life
- Human Genetics
- DNA "Fingerprinting" and Databases
- Genetic Health Testing and Discrimination
- Human Gene Alteration
- Human Cloning and Embryonic Stem Cell Research

Within each subject, entries are grouped by type (books, articles, and Internet/Web documents).

GENERAL MATERIAL ON BIOTECHNOLOGY GENETIC ENGINEERING AND GENETICS

BOOKS

Abate, Tom. *The Biotech Investor: How to Profit from the Coming Boom in Biotechnology.* New York: Times Books, 2003. Biotechnology can be a minefield for investors, but this primer is designed to help them navigate it by providing an overview of each sector of the industry and explaining where to find information and evaluate companies. Author's stance is cautious but ultimately optimistic.

Barnaby, Wendy. *The Plague Makers: The Secret World of Biological Warfare,* rev. ed. New York: Continuum Publishing Group, 2000. Describes biological warfare programs in the world's largest countries, including use of biotechnology, and the challenge of containing the biowarfare threat.

Borem, Aluizio, Fabricio R. Santos, and David E. Bowen. *Understanding Biotechnology.* Upper Saddle River, N.J.: Prentice-Hall, 2003. Introduction to the field covers history, underlying principles, key areas of research (including cloning, forensic DNA, gene therapy, and bioterrorism), and biosafety and ethical issues.

Bowring, Finn. *Science, Seeds, and Cyborgs: Biotechnology and the Appropriation of Life.* New York: Verso Books, 2003. Provides scientific and philosophical evidence to support the claim that genetic engineering's environmental, cultural, and economic risks outweigh its benefits.

Boylan, Michael, and Kevin E. Brown. *Genetic Engineering: Science and Ethics on the New Frontier.* Upper Saddle River, N.J.: Prentice-Hall, 2001. Provides background in both scientific and ethical theory and integrates the two in clear language.

Brookes, Martin. *Fly: The Unsung Hero of Twentieth-Century Science.* New York: HarperCollins/Ecco Press, 2002. Witty book uses the fruit fly, a key player in 20th-century genetic research, to provide a broad introduction to modern biology.

Brown, Stuart M. *Essentials of Medical Genomics.* Hoboken, N.J.: Wiley, 2002. Describes how genomics and proteomics, which the author calls "holistic or systems approach[es] to information flow within the cell," move beyond the study of individual genes to reveal the complex systems that make up cells, tissues, organisms, and populations. Discusses medical issues including genetic testing and gene therapy.

Campbell, J. Malcolm, and Laurie J. Heyer. *Discovering Genomics, Proteomics, and Bioinformatics.* San Francisco: Benjamin Cummings, 2002. Trains students in basic hands-on genomic analysis, including genome sequences, genome expression, and the whole genome perspective. Includes CD-ROM and integrated Web exercises.

Chadwick, Ruth, ed. *The Concise Encyclopedia of the Ethics of New Technologies.* San Diego, Calif.: Academic Press, 2001. Includes 37 articles by experts, about 10 pages each. Strong in its coverage of genetics and biotechnology.

Conley, Beverly D. *Biological Revolution.* Philadelphia: Xlibris Corp., 2002. Reviews biotechnology and related scientific developments, highlighting moral, legal, and ethical questions they raise.

Dhanda, Rahul K. *Guiding Icarus: Merging Bioethics with Corporate Interests.* Hoboken, N.J.: Wiley, 2002. Recommends that biotech companies avoid bad publicity and lawsuits by incorporating ethical considerations into their business strategies. Discusses genetically modified foods, DNA data banking, personalized medicine, and stem cell research.

Drlica, Karl. *Understanding DNA and Gene Cloning: A Guide for the Curious.* Hoboken, N.J.: Wiley, 2003. Explains the basic principles of DNA biology in a way that can be understood by college undergraduates who are not science majors.

Evans, John Hyde. *Playing God? Human Genetic Engineering and the Rationalization of Public Bioethical Debate.* Chicago: University of Chicago Press, 2002. Explores why the public has not had much to say about advancements in human genetics and genetic engineering.

Ferrari, Mauro, et al. *Opportunities in Biotechnology for Future Army Applications.* Washington, D.C.: National Academies Press, 2001. National Research Council study describes biotechnology's possible contributions to the military.

Gaisford, James D., et al., eds. *The Economics of Biotechnology.* Northampton, Mass.: Edward Elgar Publishing, 2001. Pays special attention to consumer, ethical, and environmental concerns as well as trade policy, intellectual property, and the role of international organizations such as the World Trade Organization. Focuses on agricultural biotechnology, but discussions are relevant to medical biotechnology as well.

Goldberg, Richard, and Julian Lonbay, eds. *Pharmaceutical Medicine, Biotechnology and European Law.* New York: Cambridge University Press, 2001. Distinguished legal practitioners and academics discuss European law in the areas of free movement of goods and persons, competition and intellectual property, drug regulation, biotechnology, product liability, and transnational health care litigation.

Hall, Stephen S. *Invisible Frontiers: The Race to Synthesize a Human Gene.* New York: Oxford University Press, 2002. Behind-the-scenes view of the scientific, social, and political aspects of the 1976–78 competition among three laboratories to clone a human gene for the first time, resulting in a form of the human hormone insulin that was the world's first genetically engineered drug.

Hanson, Mark J., ed. *Claiming Power over Life: Religion and Biotechnology Policy.* Washington, D.C.: Georgetown University Press, 2001. Presents eight essays explaining Christian and Jewish views of such topics as cloning and gene patenting. Claims that failure to incorporate religious views will lead to a lack of values in the making of policy on biotechnology issues.

Hargittai, Istvan, and Magdolna Hargittai, eds. *Candid Science II: Conversations with Famous Biomedical Scientists.* River Edge, N.J.: World Scientific Publishing Co., 2002. Contains 36 interviews, including 26 with Nobel laureates, bringing out the scientists' humanity as well as describing their work.

Hearns, Edwin C. *Biotechnology: A Bibliography with Indexes.* Hauppauge, N.Y.: Nova Science Publishers, 2002. Divided into sections on medicine, agriculture, biology, and business.

Howard Hughes Medical Institute. *Exploring the Biomedical Revolution.* Baltimore, Md.: Johns Hopkins University Press, 2000. Leading science writers tell the human stories behind 20th-century medical advances, including those related to genetics, from both the researchers' and the patients' points of view. Includes fold-out charts and a stereoscopic viewer that reveals 3-D images.

Keller, Evelyn Fox. *The Century of the Gene.* Cambridge, Mass.: Harvard University Press, 2002. Reconceptualizes the role of the gene, taking into account new recognition of the importance of factors outside the cell nucleus in shaping development and heredity.

King, Robert C., and William D. Stansfield. *A Dictionary of Genetics*, 6th ed. New York: Oxford University Press, 2002. Defines more than 6,500 terms and species names. Includes chronology and bibliography.

Kornberg, Arthur. *The Golden Helix: Inside Biotech Ventures.* Herndon, Va.: University Science Books, 2002. A Nobel laureate who was an advisor to the biotech company DNAX presents an insider's look at the rise and current status of the biotechnology industry. Kornberg explains why he came to feel that most productive science in the field is taking place in industry rather than academia.

Layne, Scott P., Tony J. Beugelsdijk, and C. Kumar N. Patel, eds. *Firepower in the Lab: Automation in the Fight Against Infectious Diseases and Bioterrorism.* Washington, D.C.: Joseph Henry Press, 2001. Describes new methods, some drawing on biotechnology, for combating infectious diseases, contamination of food and water, and bioterrorism by detecting and monitoring disease-causing biological agents. Material originated in a colloquium held in April 1999.

McKelvey, Maureen D. *Evolutionary Innovations: The Business of Biotechnology.* New York: Oxford University Press, 2000. Describes history of

biotechnology companies since the industry's founding, including challenges of regulation and public concern.
Moss, Lenny. *What Genes Can't Do*. Cambridge, Mass.: MIT Press, 2002. Argues that popular conceptions of the gene combine two different definitions, leading to misunderstandings. Focuses on the role of genes in cancer.
Murray, Thomas H., and Maxwell J. Mehlman, eds. *Encyclopedia of Ethical, Legal, and Policy Issues in Biotechnology*. Hoboken, N.J.: Wiley, 2000. Contains 112 in-depth, peer-reviewed articles in two volumes, covering a range of topics, including cloning, patents, and international aspects.
Nill, Kimball R. *Glossary of Biotechnology Terms*, 3d ed. Lancaster, Pa.: Technomic Publishing Co., 2002. Provides concise definitions for those outside the field, using familiar words and analogies to explain the terms.
Nossal, G. J. V., and Ross L. Coppel. *Reshaping Life: Key Issues in Genetic Engineering*, 3d ed. New York: Cambridge University Press, 2002. Concise description of the science of genetic engineering and related aspects of cell biology, as well as sociological and ethical issues related to such subjects as cloning, modified food, and gene therapy.
Oliver, Richard W. *The Biotech Age*. New York: McGraw-Hill, 2003. The author, an insider in the biotechnology industry since its beginning, discusses how the industry started, what drives it, and where he believes it is headed. He focuses on economic aspects.
Pilnick, Alison. *Genetics and Society*. Maidenhead, Berkshire, U.K.: Open University Press, 2002. Discusses the technology and social implications of prenatal genetic screening, adult genetic testing, gene therapy, genetically modified foods, and cloning.
Reilly, Philip R. *Abraham Lincoln's DNA and Other Adventures in Genetics*. Cold Spring Harbor, N.Y.: Cold Spring Harbor Laboratory Press, 2000. Wide-ranging tales of crime, history, and human behavior illustrate the principles of human genetics and the issues that genetic science raises.
Reiss, Michael J., and Roger Straughan. *Improving Nature?* New York: Cambridge University Press, 2001. Makes processes and ethical issues of genetic engineering understandable to nonbiologists. Devotes chapters to genetic modification of microorganisms, plants, animals, and humans.
Robbins-Roth, Cynthia. *From Alchemy to IPO: The Business of Biotechnology*. Cambridge, Mass.: Perseus Publishing, 2001. Optimistic view of the past and probable future of the biotechnology industry, stressing its long-term moneymaking potential.
Russell, Alan, and John Vogler, eds. *The International Politics of Biotechnology: Investigating Global Futures*. Manchester, U.K.: Manchester University Press, 2001. Essays describe biotechnology from the standpoint of international relations, discussing its relationship to the international political

economy, trade, the environment, relations between developed and developing countries, the politics of food, and biological warfare.

Schachter, Bernice. *Issues and Dilemmas of Biotechnology: A Reference Guide.* Westport, Conn.: Greenwood Publishing Group, 2000. A reference for advanced high school students and teachers that provides background on the science of biotechnology and the views of different groups about controversial issues such as genetically modified food, the patenting of human gene sequences, genetic testing, and cloning.

Sherlock, Richard, and John Morrey, eds. *Ethical Issues in Biotechnology.* Lanham, Md.: Rowman & Littlefield, 2002. Authors stress that in order to understand the ethical issues, one must first grasp the science behind the technologies. Sections cover agricultural biotechnology, food biotechnology, animal biotechnology, human genetic testing and therapy, and human cloning and stem cell research.

Steinberg, Mark L., and Sharon D. Cosloy. *The Facts On File Dictionary of Biotechnology and Genetic Engineering.* New York: Facts On File, 2000. About 2,000 entries, illustrated with drawings.

Tokar, Brian, ed. *Redesigning Life: The Worldwide Challenge to Genetic Engineering.* London: Zed Books, 2001. Scientists and activists offer reasons for opposing genetic engineering, focusing on health, food, and the environment; medical genetics and human rights; and patents, corporate power, and "biopiracy" of indigenous knowledge and resources.

Torr, James D., ed. *Genetic Engineering: Opposing Viewpoints.* San Diego, Calif.: Greenhaven Press, 2000. Anthology of articles expressing various viewpoints on questions such as how genetic engineering will affect society and how it should be regulated.

Turney, Jon. *Frankenstein's Footsteps: Science, Genetics, and Popular Culture.* New Haven, Conn.: Yale University Press, 2000. Describes how the image of science and scientists created by Mary Shelley's novel *Frankenstein* and other images from popular culture have been applied to genetics and genetic engineering and what this has to say about the way the public's views of biological science have changed over time.

Walker, Casey, ed. *Made Not Born: The Troubling World of Biotechnology.* San Francisco, Calif.: Sierra Club Books, 2000. Nineteen essays and interviews emphasize the ecological, ethical, environmental, and social dangers of genetic engineering.

Walker, John M., and Ralph Rapley, eds. *Molecular Biology and Biotechnology,* 4th ed. London: Royal Society of Chemistry, 2001. Undergraduate text with 19 chapters written by different experts. Unusually broad coverage, identifying the effects of molecular biology on biotechnology.

Witherly, Jeffre, Galen P. Perry, and Daryl L. Leja. *An A to Z of DNA Science: What Scientists Mean When They Talk About Genes and Genomes.* Cold

Spring Harbor, N.Y.: Cold Spring Harbor Laboratory Press, 2001. Defines and illustrates more than 200 specialized terms in ways that nonspecialists can understand.

Yudell, Michael, and Rob DeSalle, eds. *The Genomic Revolution: Unveiling the Unity of Life.* Washington, D.C.: National Academies Press, 2002. Seventeen papers from a September 2000 conference look at new applications of genome science in agriculture and medicine and financial, legal, and ethical issues.

Zhang, Yong-he, and Meng Zhang. *A Dictionary of Gene Technology Terms.* Boca Raton, Fla.: CRC Press-Parthenon, 2001. Some 4,500 entries cover both basic concepts and recently introduced terminology.

ARTICLES

Anderson, Clifton E. "Genetic Engineering: Dangers and Opportunities." *The Futurist,* vol. 34, March 2000, pp. 20ff. Discusses dangers and opportunities in areas including sustainable farming, dangers to the environment, and threats to genetic diversity. Follow-up article describes related trends in economics, environment, technology, society, demography (increase of lifespan), and government.

"Biotech's Yin and Yang: Chinese Biotechnology." *The Economist,* vol. 365, December 14, 2002, p. 71. China is vigorously pursuing both medical and agricultural biotechnology, but lack of infrastructure, funding, and management and the presence of confusing regulations may slow the industry's development there.

Christopher, David A. "The Gene Genie's Progeny." *World and I,* vol. 15, January 2000, pp. 172–179. Author claims that genetic engineering holds great promise for improving the health, food resources, and economic conditions of all people.

Clark, Thomas W. "Playing God, Carefully." *The Humanist,* vol. 60, May 2000, pp. 37ff. Author says that genetic engineering is too well developed for humanity to decide not to "play God," but people can decide whether to be guided by materialism or by deeper personal and social values.

"Climbing the Helical Staircase." *The Economist,* vol. 366, March 29, 2003, p. 3. Biotechnology is currently undergoing economic troubles even more severe than those besetting other high-tech industries, but insiders say that the field still has great long-term potential.

Comfort, Nathaniel C. "Are Genes Real?" *Natural History,* vol. 110, June 2001, pp. 28–37. Describes how the concept of the gene has changed in the 20th century and is still changing.

Commoner, Barry. "Unraveling the DNA Myth." *Harper's Magazine,* vol. 304, February 2002, pp. 39–47. Claims that recent discoveries in genetics, such as the Human Genome Project's finding that human beings have far fewer genes than had been thought, demonstrate that the irreducible complexity of the cell, not merely genes, controls life—and that, therefore, artificially altering genes is likely to have unexpected and potentially disastrous consequences.

Eckhardt, Robert B. "Releasing the Gene Genie." *World and I*, vol. 15, January 2000, pp. 164–171. Movement of implanted genes into species other than those in which they were placed could lead to unexpected and perhaps dangerous consequences.

Elrod-Erickson, Matthew J., and William F. Ford. "The Economic Implications of the Human Genome Project." *Business Economics*, vol. 35, October 2000, pp. 57ff. The economic effects of the famed project will likely take decades to unfold. They will show themselves in four overlapping phases: basic research, commercial investment, commercial development, and diffusion.

Epstein, Gerald L. "Bioresponsibility: Engaging the Scientific Community in Reducing the Biological Weapons Threat." *BioScience*, vol. 52, May 2002, pp. 398–399. The biological research community needs to work with law enforcement, national security, and others to reduce the threat of biological warfare by limiting access to pathogens and experimental data that can be misused.

Fairley, Peter. "Friend or Foe? Partnership with Critics Yields Results." *Chemical Week*, January 12, 2000, pp. 24ff. Only a few companies have tried to involve biotechnology critics in their decision-making processes, but sometimes the results of such partnerships have been dramatic.

Fukuyama, Francis. "Gene Regime." *Foreign Policy*, March 22, pp. 57–63. Maintains that there is presently no consistent international regulation of biotechnology, especially as it applies to humans, but development of such regulation is both possible and essential.

Gray, John. "The Unstoppable March of the Clones." *New Statesman*, vol. 131, June 24, 2002, pp. 27–29. In this author's opinion, hopes that humans will be able to control biotechnology and use it for good ends are based on "faith, not science" and are unsupported by history.

Klotzko, Arlene Judith. "Fueling the Fears of Science." *The Scientist*, vol. 16, June 24, 2002, pp. 10–11. Maintains that public fears of biotechnology are based on ignorance and impede scientific advancement.

Kwik, Gigi, et al. "Biosecurity: Responsible Stewardship of Bioscience in an Age of Catastrophic Terrorism." *Biosecurity and Bioterrorism*, vol. 1, January 2003, pp. 27–37. Urges governments and the scientific community to take steps to keep biotechnology discoveries and techniques from being used by terrorists.

Leahy, Stephen. "Biotech Hope and Hype." *Maclean's*, September 30, 2002, pp. 40–41. Author claims that economic and other benefits of biotechnology are dubious at best and that Canada should be wary of encouraging the industry.
Maienschein, Jane. "Who's in Charge of the Gene Genie?" *World and I*, vol. 15, January 2000, pp. 180ff. To regulate biotechnology effectively, people need to have intelligent discussions about competing values and establish rules to guide negotiation across differences.
Midgley, Mary. "Biotechnology and Monstrosity: Why We Should Pay Attention to the 'Yuk Factor.'" *Hastings Center Report*, vol. 30, September 2000, pp. 7ff. Leading philosopher states that emotional responses, such as the repugnance that many people feel toward some advances in biotechnology, should be examined rather than automatically dismissed; they often arise from legitimate objections.
Miller, Karen Lowry, et al. "The Biotech Boom." *Newsweek International*, October 30, 2000, pp. 56–57. Describes increased support for the biotechnology industry in several countries, including Germany, Japan, and India.
Murphy, Marina. "Terror: Future Threats: Hype About the Threat of Bioterrorism May Not Be So Far-Fetched." *Chemistry and Industry*, February 3, 2003, p. 9. Bioterrorist attacks, though not easy to pull off, are quite possible, and advances in genetic engineering make them more so.
Nelkin, Dorothy. "Beyond Risk: Reporting About Genetics in the Post-Asilomar Press." *Perspectives in Biology and Medicine*, vol. 44, Spring 2001, pp. 199ff. Examples of media coverage of gene therapy, behavioral genetics, and genetically modified foods show that the media and the public are less willing to accept scientists' judgments about their work and more focused on social and ethical aspects of genetics research than they were 25 years ago.
Rippel, Barbara. "Is There a Good and a Bad Biotechnology?" *Consumers' Research Magazine*, vol. 84, August 2001, pp. 34ff. Maintains that completely rejecting the possible contributions of biotechnology to the fight against hunger and disease is foolish; instead, advances should be evaluated individually.
"Saving the World in Comfort: Biotechnology." *The Economist*, vol. 366, March 29, 2003, p. 14. Biotechnology may have its biggest future impact as a source of environmentally friendly fuels and plastics.
Shannon, Thomas A. "The Human Genome: Now the Hard Work and the Hard Questions." *Commonweal*, vol. 128, March 23, 2001, pp. 9ff. Holds that scientists' improved understanding of the human genome should help people rediscover an ecological ethic that pictures humanity as embedded in nature rather than dominating it.

Singer, Maxine. "What Did the Asilomar Exercise Accomplish, What Did It Leave Undone?" *Perspectives on Biology and Medicine*, vol. 44, Spring 2001, pp. 186ff. A key participant claims that the legacy of 1975's famous Asilomar Conference was positive and substantial, but scientists now need to find new ways to reach the public directly.

Teitel, Martin. "Changing the Nature of Nature: An Interview with Martin Teitel." *Multinational Monitor*, January 2000, pp. 38ff. Teitel, executive director of the Council for Responsible Genetics, explains why he opposes most aspects of biotechnology and genetic engineering, including production of genetically modified food and patenting of genes and living things.

"Views from Around the World." *World Watch*, vol. 15, July–August 2002, p. 25. Scientists and officials stress the dangers of genetic engineering and say that its benefits, if any, will not extend to the world's poor.

WEB DOCUMENTS

"Bioethics Statement of Principles." BIO: Biotechnology Industry Organization. Available online. URL: http://www.bio.org/bioethics/principles.asp. Accessed on July 29, 2003. Describes the organization's position on the ethics of agricultural biotechnology, gene therapy and other issues.

"Biotechnology and Sustainability: The Fight Against Infectious Disease." Organization for Economic Co-operation and Development. Available online. URL: http://www.oecd.org/document/59/0,2340,en_2649_ 37437_2508411_1_1_1_37437,00.html. Posted on April 24, 2003. Presents case studies showing that biotechnology and genomic science can be used to fight infectious disease, particularly in the form of new vaccines, but also demonstrates that significant barriers to delivery of effective products for emerging and neglected diseases need to be overcome.

European Commission. "Life Sciences and Biotechnology—A Strategy for Europe." European Commission, Europe—The European Union On-Line. Available online. URL: http://europa.eu.int/comm/biotechnology/pdf/policypaper_en.pdf. Posted on January 23, 2002. The European Commission adopted a major policy initiative for the development of life sciences and biotechnology in Europe on January 23, 2002. This strategy paper includes an action plan with recommendations for member states, local authorities, industry, and other stakeholders.

Gaskell, George, Nick Allum, and Sally Stares. "Europeans and Biotechnology in 2002." European Commission, Europa—The European Union On-Line. Available online. URL: http://europa.eu.int/comm/public_opinion/archives/eb/ebs_177_en.pdf. Posted on March 21, 2003. "Eurobarometer" survey taken in 2002 reveals greater optimism about biotechnology, especially genetically modified foods, among Europeans than had existed before. Paper discusses values and other factors that affect feelings about biotechnology.

"Primer: Genome and Genetic Research, Patent Protection and 21st Century Medicine." BIO: Biotechnology Industry Organization. Available online. URL: http://www.bio.org/ip/primer. Accessed on July 29, 2003. Answers questions about patenting genes, genetic testing and discrimination, and gene therapy.

AGRICULTURAL BIOTECHNOLOGY AND SAFETY

BOOKS

Altieri, Miguel. *Genetic Engineering in Agriculture.* Oakland, Calif.: Food First Books, 2001. Author criticizes genetically modified crops and foods and the attempt to propagate them across the world.

Arencibia, Ariel D., ed. *Plant Genetic Engineering: Towards the Third Millennium.* St. Louis, Mo.: Elsevier, 2000. Integrates the science of plant genetic engineering with socioeconomic and environmental issues. Considers regulatory processes and public acceptance of genetically modified crops.

Atherton, Keith T., ed. *Safety Testing of Biotechnology.* Washington, D.C.: Taylor & Francis, 2001. Describes key advances in agricultural biotechnology and safety evaluation strategies applied to the industry's products, including many case studies.

Bail, Christoph, Robert Falkner, and Helen Marquard, eds. *The Cartagena Protocol on Biosafety: Reconciling Trade in Biotechnology with Environment and Development.* London: Royal Institute of International Affairs, 2003. More than 40 contributions from key stakeholders, negotiators, and analysts provide a comprehensive review of the treaty, including its scientific and commercial background, the interests at stake, the treaty's operation, and its implications and future prospects.

Bailey, Britt, and Marc Lappe, eds. *Engineering the Farm: The Social and Ethical Aspects of Agricultural Biotechnology.* Washington, D.C.: Island Press, 2002. Anthology of articles that strongly criticize the biotechnology industry and the attitudes toward social power and humans' relationship to nature that underlie it.

Boyens, Ingeborg. *Unnatural Harvest: How Genetic Engineering Is Altering Our Food.* Toronto, Ontario: Doubleday Canada, 2001. Describes agricultural biotechnology's alleged threats to biodiversity, animal welfare, and human health.

Bruce, Donald, and Ann Bruce, eds. *Engineering Genesis: The Ethics of Genetic Engineering in Nonhuman Species*, rev. ed. London: Earthscan Publications, 2000. Offers differing viewpoints on issues such as animal cloning, genetically modified food, patenting of living things, and xenotransplantation. Describes a framework for making ethical decisions.

Charles, Daniel. *Lords of the Harvest: Biotech, Big Money, and the Future of Food.* Cambridge, Mass.: Perseus Publishing, 2002. Describes the history of genetic engineering in plant crops, focusing on key individuals and groups involved in the controversy that has surrounded this technology.

Comstock, Gary L. *Vexing Nature? On the Ethical Case Against Agricultural Biotechnology.* New York: Kluwer Academic Publishers, 2000. Philosophical essays on the ethical issues raised by this technology explain how the author, originally an opponent of genetically modified crops, came to believe that many uses of these crops are morally justified.

Cummins, Ronnie, and Ben Lilliston. *Genetically Engineered Food: A Self-Defense Guide for Consumers.* New York: Marlowe & Co., 2000. Explains the dangers of genetically engineered food and what consumers can do to avoid them.

Davis, Debra. *Animal Biotechnology: Science-Based Concerns.* Washington, D.C.: National Academies Press, 2002. National Academy of Sciences report identifies science-based and policy-related concerns about animal biotechnology, including animal welfare issues and fears about the environmental effects of transgenic animals such as fast-growing salmon. Considers how animal biotechnology should be regulated.

Devries, Joseph, and Gary L. Toenniessen. *Securing the Harvest: Biotechnology, Breeding and Seed Systems for African Crops.* Cambridge, Mass.: CABI Publishing, 2002. Authors review ecological, seed supply, and policy challenges in sub-Saharan Africa, recommend priorities for research and development, and stress that biotechnology approaches to food crop improvement in the area need to consider local requirements.

Echols, Marsha A. *Food Safety and the WTO: The Interplay of Culture, Science, and Technology.* New York: Kluwer Law International, 2001. Complains that the World Trade Organization uses scientific and commercial considerations as the only guidelines in resolving disputes about food safety (including the safety of genetically modified foods), ignoring cultural, ethical, and environmental concerns and preferences.

Evenson, R. E., V. Santaniello, and D. Zilberman, eds. *Economic and Social Issues in Agricultural Biotechnology.* Cambridge, Mass.: CABI Publishing, 2002. Papers from meetings of the International Consortium on Agricultural Biotechnology Research held in 2000 and 2001 address issues such as intellectual property rights, differing needs of the public and private sectors, international market models, and expansion of biotechnology into developing countries.

Fredrickson, Donald S. *The Recombinant DNA Controversy: A Memoir, Science, Politics, and the Public Interest 1974–1981.* Washington, D.C.: American Society of Microbiology, 2001. As director of the National Institutes of Health (NIH) during this period, the author was closely involved in the early controversy about the safety of genetically modified organisms. He describes the creation and evolution of the Recombinant DNA Advisory Committee (RAC).

Houdebine, Louis-Marie. *Animal Transgenesis and Cloning.* Hoboken, N.J.: Wiley, 2003. Concise introduction to the essentials of the subject, including techniques and social and ethical implications. Covers humans as well as animals.

Hubbell, Sue. *Shrinking the Cat: Genetic Engineering before We Knew About Genes.* Boston: Mariner Books, 2002. Uses the examples of corn, the silkworm, the house cat, and the apple to show that humans have been modifying the genes of other species for thousands of years.

Isaacs, Grant E. *Agricultural Biotechnology and Transatlantic Trade: Regulatory Barriers to GM Crops.* Cambridge, Mass.: CABI Publishing, 2002. Compares the different attitudes and frameworks for risk analysis that have led to the sharp differences between the evaluations that people in the United States and Canada and people in Europe give to genetically modified crops and foods.

Kessler, Charles, and Ioannis Economidis, eds. *EC-Sponsored Research on Genetically Modified Organisms: A Review of Results.* Brussels: Research Directorate-General, European Commission, 2001. Reviews 81 EC-sponsored scientific studies and finds no evidence that genetically modified foods have harmed human health or the environment.

Lambrecht, Bill. *Dinner at the New Gene Café: How Genetic Engineering Is Changing What We Eat, How We Live, and the Global Politics of Food.* New York: St. Martin's Press, 2002. Provides a firsthand account of the controversy over genetically modified foods, focusing on industry leader Monsanto's attempt to overcome consumers' fears of these foods.

Letourneau, Deborah Kay, and Beth Elpern Burrows, eds. *Genetically Engineered Organisms: Assessing Environmental and Human Health Effects.* Boca Raton, Fla.: CRC Press, 2001. Essays on assessment of the effects of genetically engineered organisms stress that scientific understanding should guide public policy, but the people should ultimately make the decisions about safety issues.

Lurquin, Paul F. *The Green Phoenix.* New York: Columbia University Press, 2001. Tells the story behind the development of genetically engineered plants, including the field's birth as cross-fertilization between two very

different approaches to the task, its collapse and reemergence as a tool of multinational corporations, and its potentially revolutionary future.
———. *High Tech Harvest: Understanding Genetically Modified Food Plants.* Boulder, Colo.: Westview Press, 2002. Places today's agricultural biotechnology in a historical context. Focuses on scientific issues and is generally supportive of biotechnology.
Manning, Richard. *Food's Frontier: The Next Green Revolution.* Berkeley: University of California Press, 2001. Describes nine locally managed and sustainable food projects in the developing world that include use of genetic research and genetic engineering.
Martineau, Belinda. *First Fruit: The Creation of the Flavr Savr Tomato and the Birth of Biotech Food.* New York: McGraw-Hill, 2002. Provides an insider's view of the process by which the first genetically engineered food was brought to market in 1994, showing interactions among scientists, businesspeople, and government regulators.
McHughen, Alan. *Pandora's Picnic Basket: The Potential and Hazards of Genetically Modified Foods.* New York: Oxford University Press, 2000. Places genetic engineering technology in the contexts of food and environmental safety, risk assessment, corporate operations, ethics, and politics. Takes a moderate position.
Miller, Norman, ed. *Environmental Politics Casebook: Genetically Modified Foods.* Boca Raton, Fla.: Lewis Publishers, 2001. Includes documents from a variety of sources and viewpoints. This book is a companion to the author's environmental politics textbook, which stresses that public policy is formed by reconciling competing interests.
National Research Council. *Environmental Effects of Transgenic Plants: The Scope and Adequacy of Regulation.* Washington, D.C.: National Academies Press, 2002. Covers experience with existing transgenic crops, comparison of transgenic crops and those developed through conventional breeding, principles of risk assessment and management, and the science behind current regulatory frameworks. Evaluates USDA regulation and recommends more rigorous premarket review, especially of new crops that produce pharmaceutical or industrial products, and postmarket review as well.
———. *Genetically Modified Pest-Protected Plants: Science and Regulation.* Washington, D.C.: National Academies Press, 2000. Evaluates EPA regulation of pesticide Bt-containing plants. Concludes that there is no evidence that these plants are a threat to human health but stresses the urgency of establishing an appropriate framework for regulating them.
———. *Marine Biotechnology in the Twenty-First Century: Problems, Promise, and Products.* Washington, D.C.: National Academies Press, 2003. Focuses on biomedical applications of marine-based products. Considers

such subjects as drug discovery and development, applications of genomics and proteomics to marine biotechnology, and public policy.

Nelson, Gerald C., ed. *Genetically Modified Organisms in Agriculture: Economics and Politics.* San Diego, Calif.: Academic Press, 2001. Provides overview of and differing opinions about the science, politics, and economics of genetically modified organisms in agriculture, focusing on Bt corn and cotton and glyphosate-resistant soybeans. Topics discussed include regulation and international trade.

Nestle, Marion. *Safe Food: Bacteria, Biotechnology, and Bioterrorism.* Berkeley: University of California Press, 2003. Author maintains that ensuring the safety of food from accidental or deliberate contamination or from unintended effects of genetic engineering is more a matter of politics than of science and that food safety regulations need to be revised and strengthened.

Nottingham, Stephen. *Genescapes: The Ecology of Genetic Engineering.* London: Zed Books, 2002. Explains the ecological principles that underlie assessment of environmental impacts of genetically modified organisms. Stresses the technology's risks and describes safer alternatives.

Office of Technology Assessment. *Impacts of Applied Genetics: Micro-organisms, Plants, and Animals.* Washington, D.C.: Books for Business, 2002. Discusses issues including risks of genetic engineering, the patenting of living things, and public involvement in decision making about biotechnology regulation.

Paarlberg, Robert. *The Politics of Precaution: Genetically Modified Crops in Developing Countries.* Baltimore, Md.: Johns Hopkins Press, 2001. Focuses on government resistance to genetically engineered crops in Kenya, Brazil, India, and China.

Pardey, Philip G., ed. *The Future of Food: Biotechnology Markets and Policies in an International Setting.* Washington, D.C.: International Food Policy Research Institute, 2001. Essays consider biotechnology's potential role in assuring affordable and sustainably grown food for the world's population, particularly in developing countries. Discusses intellectual property rights, trade, control of research, and other economic issues.

Pence, Gregory E. *Designer Food.* Lanham, Md.: Rowman & Littlefield, 2002. Describes the "designer food" issue from the perspectives of naturalism, scientific progressivism, egalitarianism, and libertarian globalism. Sides strongly with agricultural biotechnology and opposes its critics.

Persley, Gabrielle J., and L. Reginald Macintyre, eds. *Agricultural Biotechnology: Country Case Studies.* Cambridge, Mass.: CABI Publishing, 2002. Presents 12 case studies showing the growth of agricultural biotechnology research in developing countries, as well as an overview examining issues of sustainable development, food security, and commercial applications.

Priest, Susanna Hornig. *A Grain of Truth.* Lanham, Md.: Rowman & Littlefield, 2000. Describes media coverage of the debate about genetically modified organisms and foods. Claims that the biotechnology industry has been too quick to dismiss the public's criticisms as simply the product of ignorance and sensationalist reporting.

Pringle, Peter. *Food, Inc.: Mendel to Monsanto—The Promise and Perils of the Biotech Harvest.* New York: Simon & Schuster, 2003. Claims that genetically engineered foods have great potential to reduce hunger and provide other benefits, but large multinational corporations have mismanaged them and have thereby lost the world's good will.

Qaim, Matin, Anatole F. Krattiger, and Joachim von Braun, eds. *Agricultural Biotechnology in Developing Countries: Towards Optimizing the Benefits for the Poor.* New York: Kluwer Academic Publishers, 2000. Discusses national and international constraints that limit poor people's access to the potential benefits of agricultural biotechnology in Latin America, Africa, and Asia. Urges partnerships between public and private research sectors.

Ruse, Michael, and David Castle, eds. *Genetically Modified Foods: Debating Biotechnology.* Loughton, Essex, U.K.: Prometheus Books, 2002. Presents 35 articles organized into 10 sections, covering such issues as risk assessment, labeling, developing countries, and the golden rice debate.

Santaniello, V., R. E. Evenson, and D. Zilberman, eds. *Market Development for Genetically Modified Foods.* Cambridge, Mass.: CABI Publishing, 2002. Papers from the fourth meeting of the International Consortium on Agricultural Biotechnology Research, held in August 2000, are divided into sections focusing on consumer reactions to information about genetically modified foods, regulatory issues, farmers' acceptance of biotechnology products, and changes in industrial organization in the food and life science sectors.

Shiva, Vandana. *Tomorrow's Biodiversity.* London: Thames and Hudson, 2001. Criticizes genetically engineered crops as one means by which multinational corporations are spreading monoculture and reducing biodiversity worldwide, thereby threatening the environment and food security.

Swanson, Timothy M., ed. *Biotechnology Agriculture and the Developing World: The Distributional Implications of Technological Change.* Northampton, Mass.: Edward Elgar Publishing, 2002. Eleven contributions based on research by Britain's Department for International Development investigate ways in which changes brought by agricultural biotechnology are likely to affect global diffusion of the technology, especially in the developing world.

Teitel, Martin, and Kimberly A. Wilson. *Genetically Engineered Food: Changing the Nature of Nature,* 2nd ed. Rochester, Vt.: Inner Traditions International, 2001. Describes three alleged problems with genetically engineered food: future uncertainties, ownership of seeds, and globalization of monoculture. Stresses risks of the technology and tells consumers how to avoid its products.

Wambugu, Florence. *Modifying Africa.* Red Hills, Hyderabad, India: Pragati Offset Pvt. Ltd., 2001. A Kenyan scientist trained in England and the United States, involved in developing a genetically engineered yam (a major food crop in sub-Saharan Africa) that is able to resist an important viral pest, praises the possible benefits of biotechnology for feeding the hungry continent.

Winston, Mark L. *Travels in the Genetically Modified Zone.* Cambridge, Mass.: Harvard University Press, 2002. Includes interviews with farmers, activists, industrial workers, and publicists. Discusses issues such as patents and the control of seeds, consumer and environmental safety, and agribusiness profits; suggests solutions for controversial problems.

ARTICLES

Ackerman, Jennifer. "Food: How Altered?" *National Geographic*, vol. 201, May 2002, pp. 33–50. Describes the extent of agricultural genetic engineering and discusses arguments for and against the technology.

Altieri, Miguel A. "No: Poor Farmers Won't Reap the Benefits." *Foreign Policy*, Summer 2000, pp. 123ff. The world's poor farmers are unlikely to reap the benefits of genetically engineered crops because they cannot afford the expensive, patented seeds, author maintains; such crops may also decrease the diversity of local ecosystems by wiping out native insects and plants, thus further threatening food security.

Bjerklie, David, Dan Cray, and Dick Thompson. "Monkey Business." *Time*, vol. 157, January 22, 2001, pp. 40–41. Describes the creation of the first transgenic primate, a rhesus monkey named ANDi (for "inserted DNA" backward), whose cells contain a jellyfish gene as a marker.

Bren, Linda. "Cloning: Revolution or Evolution in Animal Production?" *FDA Consumer*, vol. 37, May–June 2003, pp. 28–33. Cloned cattle and other domestic animals are becoming more common, but concerns about them remain, and the Food and Drug Administration is still deciding whether products from them are suitable for human consumption.

Brown, Kathryn S., "Food with Attitude." *Discover*, vol. 21, March 2000, pp. 31–32. Discusses edible vaccines and other "functional foods," altered by conventional breeding or genetic engineering to contain enhanced health benefits.

Capron, Alexander M., and Renie Schapiro. "Remember Asilomar?" *Perspectives in Biology and Medicine*, vol. 44, Spring 2001, pp. 162ff. On the 25th anniversary of the historic Asilomar conference, in which scientists voluntarily agreed to postpone certain recombinant DNA experiments they deemed to be potentially dangerous, scientists and others met to evaluate the conference and its legacy and consider how its ideals apply to genetic research today.

Carey, John. "Are Bio-Foods Safe?" *Business Week*, December 20, 1999, pp. 70ff. Presents safety issues related to genetically modified food crops and describes arguments on both sides.

Charles, Daniel. "Corn That Clones Itself." *Technology Review*, vol. 106, March 2003, pp. 32–38. Apomixis, a natural process that allows characteristics to reproduce dependably, might be introduced into genetically engineered corn and other crops, but both critics of GM crops and the multinational companies that make the crops oppose the idea because it would allow farmers to reuse altered seeds from year to year.

Clapp, Stephen. "Biotech Companies Join Anti-Hunger Effort in Africa." *Pesticide and Toxic Chemical News*, vol. 31, March 17, 2003, p. 6. Several large agricultural biotechnology companies, including Monsanto, plan to work with the newly formed African Agricultural Technology Foundation to reduce hunger and poverty in Africa through introduction of genetically modified crops.

———. "North America, Europe Differ Radically on Biotech Regulation." *Food Chemical News*, vol. 44, February 18, 2002, p. 5. Summarizes differences between American and European frameworks for analyzing the risks of biotechnology. The former focuses on "scientific rationality," whereas the latter stresses "social rationality."

———. "Traceability: A Global Perspective." *Food Traceability Report*, vol. 2, May 2002, pp. 16–17. "Traceability" of genetically modified elements in food is becoming the new watchword of European Union regulation, but the proposed standards satisfy neither European environmentalists nor American industrialists.

Clarke, Paul. "Second Thoughts." *E*, vol. 12, November–December 2001, p. 14. Claims that American farmers, fearful of losing markets and damaging the environment, are losing their enthusiasm for genetically modified crops.

De Greef, Willy. "Challenging the Food Crisis: Is There a Place for Biotechnology in Agriculture?" *OECD Observer*, Summer 2000, p. 84. Holds that eradicating hunger requires improvement in both production (quantity) and distribution of food and that biotechnology can help with the former.

Demenet, Philip. "Can Genetically Modified Organisms Feed the World?" *UNESCO Courier*, September 2001, pp. 10ff. Holds that the world's poor need other improvements, such as better soil management, more than they need genetically modified crops, which may threaten biodiversity.

"EU Unveils GMO Traceability and Labeling Rules." *Agra Europe*, July 27, 2001, p. EP/1. Article summarizes the new rules, which drew criticism from both environmental groups and U.S. commercial interests.

Evenson, Christian, Thomas Hoban, and Eric Woodrum. "Technology and Morality: Influences on Public Attitudes Toward Biotechnology." *Knowledge Technology and Policy*, vol. 13, Spring 2000, pp. 43ff. Describes effects of gender, religious belief, and perceptions of personal benefit and environmental risk on attitudes toward agricultural biotechnology. Concludes that education and increased awareness of the new technology, especially when it is presented in a positive light, are likely to decrease objections to it.

"Feeding the Five Billion: New Agricultural Techniques Can Keep Hunger at Bay." *The Economist*, November 10, 2001, pp. 4ff. Says that well-fed but suspicious citizens in Europe and elsewhere, who fear unquantified risks from genetically modified foods, may keep the world's poor and hungry from benefiting from the increased productivity promised by these improved crops.

Fleeson, Lucinda. "A Cure for the Common Farm?" *Mother Jones*, vol. 28, March–April 2003, pp. 17–19. Claims that crops genetically engineered to produce drugs could help small farms because they do not need to be grown in large quantities in order to be profitable, but they may also present dangers.

"Genetically Engineered Foods." *Nutrition Action Healthletter*, vol. 28, November 2001, pp. 1ff. In this interview, Doug Gurian-Sherman and Gregory Jaffe, codirectors of the Biotechnology Project at the Center for Science in the Public Interest, call for stronger government regulation of GM crops but are cautiously optimistic about their use.

Gepts, P. "A Comparison between Crop Domestication, Classical Plant Breeding, and Genetic Engineering." *Crop Science*, vol. 42, November–December 2002, pp. 1780–1790. Claims that uses of agricultural genetic engineering need to be evaluated on a case-by-case basis and that the technique is likely to increase in precision, usefulness to consumers, and friendliness to the environment.

Goklany, Indur M. "The Future of Food." *Forum for Applied Research and Public Policy*, vol. 16, Summer 2001, pp. 59ff. Proposes a framework for applying the "precautionary principle" to evaluate policies for regulating genetically modified crops. Author concludes that potential benefits of these crops outweigh potential harms in both likelihood and scale and that development of GM crops therefore should proceed, although cautiously.

"Goodbye Dolly . . . and Friends?" *Lancet*, vol. 361, March 1, 2003, p. 711. Although their being clones may not have caused their demise, the recent deaths of Scotland's famous Dolly and an Australian sheep clone, Matilda, highlight serious weaknesses in reproductive cloning of animals.

Guterl, Fred, et al. "Brave New Foods." *Newsweek International*, January 28, 2002, p. 46. Plants may be genetically engineered to produce vaccines, but they face an uphill battle for approval and funding unless public fears of genetically modified foods are eased.

Hart, Kathleen. "Advisory Group Urges USDA to Endorse GM Thresholds for Non-GM Foods." *Food Chemical News*, vol. 43, April 30, 2001, p. 3. Members of the USDA's Advisory Committee on Agricultural Biotechnology have urged the department to endorse the concept of thresholds for accidental GM contamination of non-GM foods because they believe that some contamination is unavoidable.

Ho, Mae-Wan. "Horizontal Gene Transfer Happens." *Synthesis/Regeneration*, Winter 2001, pp. 45ff. Claims that movement of altered genes from one species to another does occur and that bacteria and viruses can accelerate it.

Holdrege, Greg, and Steve Talbott. "Sowing Technology." *Sierra*, vol. 86, July 2001, pp. 34ff. Provides ecological arguments against high-tech agriculture in general and genetically engineered crops in particular.

Howie, Michael. "'No Debate' That Africa Needs Biotechnology." *Feedstuffs*, vol. 73, July 9, 2001, p. 1. Two African experts stress that the continent needs agricultural biotechnology; they criticize those who oppose its use.

Hunter, Beatrice Trum. "Biotech Animals Come to the Farm." *Consumers' Research Magazine*, vol. 84, May 2001, pp. 24–25. Fast-growing salmon and other genetically engineered animals offer potential threats as well as benefits. The FDA currently classifies and regulates such animals as drugs; opinions differ as to whether this agency and its regulatory pattern are adequate to protect the environment and consumers.

Jordan, Carl F. "Genetic Engineering, the Farm Crisis, and World Hunger." *BioScience*, vol. 52, June 2002, pp. 523–529. Maintains that agricultural biotechnology is unlikely to help farmers in the United States or many other countries because overproduction is the cause of the farmers' economic hard times.

Kaplinsky, Nicholas, et al. "Maize Transgene Results in Mexico Are Artefacts." *Nature*, vol. 416, April 11, 2002, pp. 601–602. One of several refutations of claim by David Quist and Ignacio Chapela in an earlier issue of *Nature* that genes from genetically engineered corn had appeared in Mexican native corn.

Kearns, Peter. "GM Food: Science, Safety, and Society." *OECD Observer*, September 2001, pp. 41–42. The Organization for Economic Cooperation and Development (OECD) describes concerns about genetically modified foods and explains steps the organization is taking to meet public demands to know how decisions about such foods are reached and to be involved in those decisions.

Kneen, Brewster. "Caring for Life: Genetic Engineering and Agriculture." *Ecumenical Review*, vol. 54, July 2002, pp. 262–270. Claims that those who say that agricultural biotechnology will provide food security are engaging in an unquestioning idolatry of technology, attributing miraculous redemptive powers to it.

Kriz, Margaret. "Global Food Fight." *National Journal*, vol. 32, March 4, 2000, pp. 688ff. Disputes about genetically modified crops leave American farmers caught in the middle, unsure of what to plant.

Lavendel, Brian. "Jurassic Ark." *Animals*, vol. 134, Summer 2001, pp. 16–17. Cloning extinct or endangered species may be possible—but many conservationists doubt that it is a good approach to species preservation.

Leisinger, Klaus M. "Yes: Stop Blocking Progress." *Foreign Policy*, Summer 2000, pp. 113ff. The author, executive director of the Novartis Foundation for Sustainable Development, maintains that the potential benefits of genetically engineered crops outweigh their risks.

Lilliston, Ben. "Don't Ask, Don't Know: The Biotech Regulatory Vacuum." *Multinational Monitor*, January 2000, pp. 9ff. Increasing numbers of critics say that U.S. regulatory agencies have allowed genetically engineered crops to be introduced into the environment and the food supply without sufficient safety testing.

Mann, Charles C. "Has GM Corn 'Invaded' Mexico?" *Science*, vol. 295, March 1, 2002, pp. 1617–1618. Scientists are arguing violently over a paper published in the November 29, 2001, issue of the respected science journal *Nature*, which claims that genes from genetically altered corn have appeared in native corn in Mexico and that the genes are unstable, possibly creating unpredictable effects.

Manning, I. Richard. "Eating the Genes." *Technology Review*, vol. 104, July 2001, p. 90. India has been using biotechnology to produce crops resistant to diseases and pests, yet millions there are still hungry.

McGloughlin, Martina. "Biotechnology Offers Solution to World Food Dilemma." *Journal of the American Dietetic Association*, vol. 100, October 2000, p. 1. Stresses the benefits and safety of genetically altered crops and challenges what the author calls misconceptions of biotech critics.

Miller, Henry I. "EPA Disregards Science: New Biotech Rules Not Based on Risk." *Consumers' Research Magazine*, vol. 84, August 2001, p. 26. Opposes EPA regulations that treat Bt-containing crops like chemical pesticides. Praises the crops for reducing pesticide use.

———. "Food Biotechnology: A Microcosm of Science Under Attack." *Knowledge Technology and Policy*, vol. 13, Spring 2000, p. 39. Claims that efforts of "antitechnology extremists" to stop use of genetically modified foods threaten research that could create safer, cheaper, and more nutritious foods.

Miller, Henry I., and Gregory Conko. "The Science of Biotechnology Meets the Politics of Global Regulation." *Issues in Science and Technology*, vol. 17, Fall 2000, pp. 47ff. Opposes new, stricter regulation of genetically modified crops, claiming that demands for it are based on politics and environmental fears rather than on scientific evidence.

Mittal, Anuradha, and Peter Rosset. "Genetic Engineering and the Privatization of Seeds." *Dollars and Sense*, March 2001, pp. 24ff. Describes protests in India and other developing countries against Monsanto and other giants of agricultural biotechnology. Claims that agricultural biotechnology will not benefit the developing world unless it can be made to help poor rather than wealthy farmers.

Moore, Julia A. "More Than a Food Fight." *Issues in Science and Technology*, vol. 17, Summer 2001, pp. 31ff. Maintains that disagreements about agricultural biotechnology between the United States and Europe reveal the dangers of a lack of dialogue between scientists and the public.

Nanda, Ved P. "Genetically Modified Food and International Law—the Biosafety Protocol and Regulations in Europe." *Denver Journal of International Law and Policy*, vol. 28, Summer 2000, pp. 235ff. Analyzes regulation of genetically modified organisms in Europe and the Biosafety Protocol, which attempts to regulate international trade in such organisms. Concludes that European regulations are the more effective of the two and recommends creation of a new world body to monitor biotechnology.

Nash, J. Madeleine. "Grains of Hope." *Time International*, vol. 157, February 12, 2001, pp. 34ff. Focuses on "golden rice," which potentially prevents or cures vitamin A deficiency, as an example of genetically engineered crops that may reduce world hunger and malnutrition, but also points out the crops' environmental risks.

O'Reilly, Brian. "Reaping a Biotech Blunder." *Fortune*, vol. 143, February 19, 2001, pp. 156ff. Describes the fall 2000 incident in which StarLink, a type of genetically engineered corn approved for animal but not human food because it might cause allergic reactions, was found to contaminate human food products, resulting in a massive recall. Although no one was provably harmed by the corn, O'Reilly calls the affair "a disturbingly close brush with disaster . . . and a wake-up call."

Paarlberg, Robert L. "Reinvigorating Genetically Modified Crops." *Issues in Science and Technology*, vol. 19, Spring 2003, pp. 86–92. Maintains that the poor and hungry in developing nations will be the greatest losers if strict European-style restrictions on genetically modified crops are adopted in other parts of the world.

Palevitz, Barry A. "Corn Goes Pop, then Kaboom." *The Scientist*, vol. 16, April 29, 2002, pp. 18–19. Explains how in April 2002, the respected science journal *Nature* took the unusual step of apologizing for its November 2001 printing of a paper stating that genes from genetically altered corn had been found in Mexican corn plants. This was the culmination of the bitter controversy that had surrounded the paper.

Phifer, P. R., and Wolfenbarger, LaReesa. "The Ecological Risks and Benefits of Genetically Engineered Plants." *Science*, vol. 290, December 15, 2000, pp. 2088ff. Stresses that key experiments to establish both risks and benefits of genetically engineered crops have yet to be performed and that it is important to judge these crops by field studies, where unexpected events are more likely to occur than they are under controlled laboratory conditions.

Pollack, Mark A., and Gregory C. Shaffer. "Biotechnology: The Next Transatlantic Trade War?" *Washington Quarterly*, vol. 23, Autumn 2000, pp. 41ff. Concludes that the conflict between the United States and Europe over regulation of genetically modified organisms is rooted in longstanding, opposing philosophies of safety regulation, but the dispute is unlikely to escalate into a full-scale trade war.

Powell, Douglas. "Five Reasons Why GMO Food Labels Don't Work." *National Post*, June 22, 2001. Among other things, Powell claims that it is hard (if not impossible) to create food that is truly free of genetically modified material and that genetic modification is hard to define precisely, leaving labeling subject to fraud.

Quist, David, and Ignacio H. Chapela. "Transgenic DNA Introgressed into Traditional Maize Landraces in Oaxaca, Mexico," *Nature*, vol. 414, November 29, 2001, pp. 541–543. Controversial article claims to have found genes from genetically engineered corn in Mexican native corn, even though Mexico does not allow planting of genetically altered corn; this suggests that gene flow from engineered crops has occurred.

Robbins, John. "A Biological Apocalypse Averted." *Earth Island Journal*, vol. 16, Winter 2001, p. 27. Claims that a genetically engineered soil bacterium, intended to break down organic matter and create ethanol for fuel, destroyed vital fungi in soil and might have ended all plant life in North America if it had been released.

Sardar, Ziauddin. "Thank You for the Genes We Eat." *New Statesman*, vol. 130, January 15, 2001, pp. 28–29. A skeptical reporter finds close relationships between U.S. regulatory agencies and the biotechnology industry and a shared faith in the benefits of science in general and genetically modified crops in particular.

Sharratt, Lucy. "Building Resistance in Canada: A View from the North." *Synthesis/Regeneration*, Fall 2000, pp. 39ff. States that disapproval of genetic engineering, particularly of food crops, is growing in Canada.

Snell, Marilyn Berlin. "Against the Grain." *Sierra*, vol. 86, July 2001, pp. 30ff. Interview with Tewolde Berhan Gebre Egziabher, former head of Ethiopia's Environment Protection Authority. He opposes introduction of transgenic crops into the developing world because of the technology's control by large corporations and the risk that it will damage the environment.

Specter, Michael. "The Pharmageddon Riddle." *New Yorker*, vol. 76, April 10, 2000, pp. 58ff. Describes how the controversy over genetically modified foods has affected biotechnology corporate giant Monsanto.

Stone, Glenn Davis, et al. "Fallacies in the Genetic-Modification Wars, Implications for Developing Countries, and Anthropological Perspectives." *Current Anthropology*, vol. 43, August–October 2002, pp. 611–630. Concludes that the question of whether genetically engineered crops can help feed the hungry in developing countries is much more complex than either proponents or opponents admit.

Strauss, Mark. "When Malthus Meets Mendel." *Foreign Policy*, Summer 2000, pp. 105ff. Claims that as the United States and Europe argue over the safety of genetically modified crops and foods, the hungry people of the developing world, who potentially have the most to gain from the new technology, are forced to remain on the sidelines of the dispute.

Tait, Malcolm. "Bessie and the Gaur." *The Ecologist*, vol. 30, December 2000, p. 46. Maintains that cloning of endangered or extinct species distracts attention from legitimate wildlife conservation and is merely the latest example of humans' attempts to dominate nature.

Tangley, Laura. "Engineering the Harvest." *U.S. News & World Report*, vol. 128, March 13, 2000, p. 46. Concludes that the developing world has a pressing need for genetically modified crops.

Taylor, Sarah E., and John F. C. Luedke. "FDA Approval Process Ensures Biotech Safety." *Journal of the American Dietetic Association*, vol. 100, October 2000, p. 3. Reviews current and proposed procedures for FDA safety review of genetically modified foods.

Thomassin, Paul J., and L. Martin Clautier. "Informational Requirements and the Regulatory Process of Agricultural Biotechnology." *Journal of Economic Issues*, vol. 35, June 2001, pp. 323ff. Investigates the reinforcing and balancing regulatory pressures that affect agricultural and food biotechnology in Canada.

Thompson, Larry. "Are Bioengineered Foods Safe?" *FDA Consumer*, vol. 34, January 2000, pp. 18ff. Interview with FDA commissioner Jane E. Henney, who reassures consumers about the agency's regulation process and the safety of genetically modified foods.

Tokar, Brian. "Biohazards: The Next Generation?" *Synthesis/Regeneration*, Summer 2001, p. 37. Claims that plants bioengineered to produce drugs and industrial chemicals, now under development, are likely to offer much greater threats to health and the environment than present genetically modified crops.

———. "Butterfly Experiment Highlights Biotech Hazards." *Synthesis/Regeneration*, Winter 2000, pp. 28ff. States that Cornell University research showing that pollen from corn engineered to produce Bt pesticide kills monarch butterfly caterpillars is an example of proofs that bioengineered crops present threats to the environment.

Van Reenen, C. G., et al. "Transgenesis May Affect Farm Animal Welfare." *Journal of Animal Science*, vol. 79, July 2001, pp. 1763ff. Considers possible ways in which genetic engineering of animals may harm their welfare and discusses strategies for evaluating the health and welfare of transgenic farm animals.

Victor, David G., and C. Ford Runge. "Farming the Genetic Frontier." *Foreign Affairs*, vol. 81, May–June 2002, pp. 107ff. Claims that detractors will succeed in blinding people to agricultural biotechnology's potential benefits, especially to the poor, unless the United States and other countries pursue a long-term strategy for managing and regulating genetically engineered crops. Lists three elements that such a strategy must have.

Weiner, Charles. "Drawing the Line in Genetic Engineering." *Perspectives in Biology and Medicine*, vol. 44, Spring 2001, pp. 208ff. Author concludes that scientists' self-regulation was effective in the early days of recombinant DNA technology, but it is no longer adequate. Decisions about genetic engineering now must be made by the public and take into account social justice and moral values as well as science-related issues.

Wheelwright, Jeff. "Don't Eat Again Until You Read This." *Discover*, vol. 22, March 2001, pp. 38–42. Describes the controversy that arose in fall 2000, when StarLink, a type of genetically engineered corn approved for animal feed but not for human food because it might cause allergic reactions, was found to have contaminated a wide array of corn-containing food products, producing a massive recall.

Wright, Karen. "Terminator Genes." *Discover*, vol. 24, August 2003, pp. 49–51. Inserted genes that keep crop plants from reproducing, thereby forcing farmers to buy new seeds each year rather than replanting from previous crops as they traditionally do, have drawn widespread criticism, but agribusiness companies are still experimenting with them.

Wright, Susan. "Legitimating Genetic Engineering." *Perspectives in Biology and Medicine*, vol. 44, Spring 2001, pp. 235ff. Claims that scientists at the historic 1975 Asilomar conference were motivated chiefly by desire to persuade the public and legislators to allow them to set their own rules for ensuring the safety of recombinant DNA experiments.

WEB DOCUMENTS

"Agricultural Biotechnology: Myths and Facts." IFIC (International Food Information Council) Foundation. Available online. URL: http://ific.org/publications/other/biotechmythsom.cfm. Posted in October 2001. Rebuts arguments and criticisms from opponents of agricultural biotechnology and genetically modified foods, which the organization calls myths.

Ammann, Klaus. "Biodiversity and Agricultural Biotechnology: A Review of the Impact of Agricultural Biotechnology on Biodiversity." Bio-scope. Available online. URL: http://www.bio-scope.org/attach/debates/Report-Biodiv-Biotech3.pdf. Accessed on July 30, 2003. Concludes that the greatest threats to biodiversity are habitat loss and fragmentation and that GM crops can minimize these by increasing productivity of existing farmland. Genetic modification does not represent any greater risk to crop genetic diversity than standard breeding programs.

"Animals and Biotechnology." Agriculture and Environment Biotechnology Commission (AEBC). Available online. URL: http://www.aebc.gov.uk/aebc/pdf/animals_and_biotechnology_report.pdf. Posted in September 2002. Report from agency advising the British government on biotechnology issues. Its recommendations include establishing a strategic advisory body to examine the issues raised by application of genetic engineering to farm animals and involving the public in decision making.

Benbrook, Charles M. "The Farm-Level Economic Impacts of *Bt* Corn from 1996 through 2001: An Independent National Assessment." Ag BioTech InfoNet. Available online. URL: http://www.biotech-info.net/GMO_corn.pdf. Posted in December 2001. Technical report evaluating whether yield benefits have justified the high price that U.S. farmers pay for pest-resistant (Bt-containing) genetically modified corn. Concludes that gains are marginal at best and that pests can be dealt with equally effectively in other ways.

"Biodiversity: The Impact of Biotechnology." European Federation of Biotechnology. Available online. URL: http://www.efbpublic.org/site/page.liones.php?pointer=1-2-20-123-608. Posted on January 25, 2002. Concludes that biotechnology has the potential to improve sustainability in several ways and can help to maintain both natural and agricultural biodiversity.

"Biotechnology and Foods." Institute of Food Technologists. Available online. URL: http://www.ift.org/cms/?pid-1000380. Posted on September 19, 2000. This review of scientific evidence related to biotechnology and foods discusses human food safety, labeling, and benefits and concerns. It concludes that genetically modified foods are safe and beneficial.

Caplan, Richard. "Raising Risk: Field Testing of Genetically Engineered Crops in the United States." U.S. Public Interest Research Group. Available online. URL: http://www.uspirg.org/reports/RaisingRisk2003.pdf. Posted in June 2003. Highlights potential health and environmental risks associated with the release of genetically engineered plants.

Annotated Bibliography

Conway, Gordon. "From the Green Revolution to the Biotechnology Revolution: Food for Poor People in the 21st Century." Rockefeller Foundation. Available online. URL: http://www.rockfound.org/documents/566/Conway.pdf. Posted on March 12, 2003. President of the Rockefeller Foundation explains why he feels that biotechnology can help Africa feed its people without repeating the environmental mistakes of the Green Revolution.

"Crops on Trial." Agricultural and Environment Biotechnology Commission (AEBC). Available online. URL: http://www.aebc.gov.uk/aebc/pdf/crops.pdf. Posted in September 2001. Report from agency advising the British government on biotechnology issues discusses farm-scale evaluations, the legal and regulatory framework, and public reactions to genetically modified crops.

"Crops Under Question." The GE Food Alert Campaign Center. Available online. URL: http://www.gefoodalert.org/library/admin/uploadedfiles/Crops_Under_Question_A_Briefing_Book_on_Geneti.pdf. Accessed on July 31, 2003. Focuses on EPA regulation of Bt-containing crops. Expresses concern about human allergies, damage to nonpest insects such as butterflies, and development of resistance in pest insects. Concludes that the regulation process should be made more stringent and more accessible to the public.

Druker, Steven M. "The Poor Performance of Genetically Engineered Crops." Alliance for Bio-Integrity. Available online. URL: http://www.bio-integrity.org/increased-risks-no-benefits.htm. Posted on June 3, 2003. Claims that genetically engineered crops reduce yields, lower profits for farmers, and create hazards for the environment and food safety.

———. "Why Concerns About Health Risks of Genetically Engineered Food Are Scientifically Justified." Alliance for Bio-Integrity. Available online. URL: http://www.bio-integrity.org/health-risks/health-risks-ge-foods.htm. Posted in 2001. Maintains that claims about the safety of genetically engineered foods are based on flawed assumptions and opposed by many scientists. Urges stronger federal regulation.

"Environmental Benefits: More Studies Show How Biotech Crops Help Wildlife, Environment." Council for Biotechnology Information. Available online. URL: http://www.whybiotech.com/index.asp?id=1805. Posted in 2003. Environmental benefits cited include prevention of soil erosion, increase in diversity of beneficial insects, and increased yield from marginal land.

"Environmental Savior or Saboteur? Debating the Impacts of Genetic Engineering." Pew Initiative on Food and Biotechnology. Available online. URL: http://pewagbiotech.org/research/survey1-02.pdf. Posted on February 4, 2002. Provides highlights from a survey of 1,214 adults throughout the United States, conducted in January 2002. After reading informational statements about agricultural biotechnology, those surveyed were about evenly divided about whether this technology helps or hurts the environment.

Fakuda-Parr, Sakiko, et al. "Human Development Report 2001: Making New Technologies Work for Human Development." United Nations Development Programme. Available online. URL: http://hdr.undp.org/reports/global/2001/en/. Posted in 2001. Claims that developing nations can benefit from genetically engineered crops if the crops are regulated to ensure safety and if outside critics, lack of resources and technical expertise, and other problems do not block their use.

"Field Work: Weighing up the Costs and Benefits of GM Crops." 10 Downing Street web site. Available online. URL: http://www.number10.gov.uk/files/pdf/GMreport.pdf. Posted on July 11, 2003. Report of the prime minister's Strategy Unit's analysis of the costs and benefits of cultivating genetically modified crops in Britain concludes that benefits are likely to be limited in the short term but may be greater in the long term.

"Food Biotechnology." International Food Information Council. Available online. URL: http://ific.org/food/biotechnology/index.cfm. Posted in March 2002. Overview of agricultural biotechnology and its regulation stresses the benefits of the technology.

Freese, Bill. "Manufacturing Drugs and Chemicals in Crops." The GE Food Alert Campaign Center. Available online. URL: http://www.gefoodalert.org/library/admin/uploadedfiles/Manufacturing_Drugs_and_Chemicals_in_Crops.doc. Posted in July 2002. Claims that crops genetically engineered to produce drugs and chemicals pose threats to consumers, farmers, food companies, and the environment. For example, such compounds could leach into the soil through plant roots and thereby reach other plants or the water supply.

Fresco, Louise O. "Which Road Do We Take?" Food and Agriculture Organization of the United Nations (FAO). Available online. URL: http://www.fao.org/ag/magazine/fao-gr.pdf. Posted on January 30, 2003. Claims that biotechnology investment generally ignores staple food crops of developing countries and that the "molecular divide" between developed and developing countries may exacerbate existing inequalities. Recommends a new social contract for sustainable agriculture.

"Future Fish: Issues in Science and Regulation of Transgenic Fish." Pew Initiative on Food and Biotechnology. Available online. URL: http://pewagbiotech.org/research/fish. Posted on January 14, 2003. Questions whether the FDA's plans and power to regulate transgenic fish are adequate to deal with environmental/ecological concerns raised by these and possibly other transgenic animals.

"Genetically Modified Foods." The United States General Accounting Office. Available online. URL: http://www.gao.gov/new.items/d02566.pdf. Posted in May 2002. Based on interviews with a variety of experts both within and outside the industry, concludes that the FDA's safety reviews of such foods are adequate but could be improved to give consumers more confidence that genetically modified foods are safe to eat.

Gianessi, Leonard P., et al. "Plant Biotechnology: Current and Potential Impact for Improving Pest Management in U.S. Agriculture: An Analysis of 40 Case Studies." National Center for Food and Agricultural Policy. Available online. URL: http://www.ncfap.org/40CaseStudies/MainReport.pdf. Posted in June 2002. Claims that genetically engineered crops increase production, save farmers money, and substantially reduce pesticide use.

"GM Science Review: First Report." GM Science Review. Available online. URL:http://www.gmsciencedebate.org.uk/report/pdf/gmsci-report1-full.pdf. Posted in July 2003. Panel of experts recommends evaluating genetic engineering applications on a case-by-case basis rather than either accepting or rejecting the technology as a whole. Emphasizes need for careful regulation, particularly regarding effects on farmland and wildlife.

"Guide to U.S. Regulation of Genetically Modified Food and Agricultural Biotechnology Products." Pew Initiative on Food and Biotechnology. Available online. URL: http://pewagbiotech.org/resources/ issuebriefs/1-regguide.pdf. Accessed on August 8, 2003. Includes evolution of biotechnology regulation, regulation by type of organism, and regulation of food, drugs, and other products derived from transgenic organisms.

Gurian-Sherman, Doug. "Holes in the Biotech Safety Net." Center for Science in the Public Interest. Available online. URL: http://cspinet.org/new/pdf/fda_report__final.pdf. Posted on January 7, 2003. Claims that FDA regulatory policy does not assure the safety of genetically modified foods and makes recommendations for tightening the regulations.

"Harvest on the Horizon: Future Uses of Agricultural Biotechnology." Pew Initiative on Food and Biotechnology. Available online. URL: http://pewagbiotech.org/research/harvest/harvest.pdf. Posted in September 2001. Covers food crops, trees, turfgrass, and flowers, as well as genetic engineering for production of pharmaceuticals, industrial products, and bioremediation.

Hickey, Ellen, and Anuradha Mittal, eds. "Voices from the South: The Third World Debunks Corporate Myths on Genetically Engineered Crops." Food First/Institute for Food and Development Policy. Available online. URL: http://www.foodfirst.org/sacramento/voices/voicesfull.pdf.

Posted in May 2003. A collection of criticisms of genetically engineered crops from the developing world, focusing on the crops' control by large companies and threats to farmers' right to replant seeds.

Ho, Mae-Wan, and Lim Li Ching. "The Case for a GM-Free Sustainable World." Food First/Institute for Food and Development Policy. Available online. URL: http://www.foodfirst.org/progs/global/ge/isp/ispreport.pdf. Posted on June 15, 2003. This report from an independent science panel stresses the problems and hazards of genetically modified crops, the inadequacy of government testing, and the benefits of sustainable agriculture that does not use these crops.

Innovest Strategic Value Advisors. "Monsanto and Genetic Engineering: Risks for Investors." Greenpeace USA. Available online. URL: http://www.greenpeaceusa.org/images/user/2/i237.pdf. Posted in April 2003. Explains why Innovest Strategic Value Advisors believes that Monsanto is a bad investment and that its genetically engineered crops are dangerous to health and the environment.

McGarity, Thomas O., and Patricia I. Hansen. "Breeding Distrust: An Assessment and Recommendations for Improving the Regulation of Plant Derived Genetically Modified Foods." Ag BioTech InfoNet. Available online. URL: http://www.biotech-info.net/Breeding_Distrust.html. Posted on January 11, 2001. Considers risks and benefits, describes current policy and regulation in the United States and elsewhere, lists elements of what authors regard as an adequate regulatory regime, and recommends improvements in U.S. regulatory policy.

Mellon, Margaret, and Jane Rissler. "Environmental Effects of Genetically Modified Crops: Recent Experiences." Union of Concerned Scientists. Available online. URL: http://www.ucsusa.org/food_and_environment/biotechnology/page.cfm? pageID=1219. Posted on June 13, 2003. Concludes that the scientific evidence available to date, while encouraging, does not support the conclusion that genetically modified crops are safe for human health or the environment. Warns that the next generation of such crops, which will produce drugs and industrial chemicals, may be more dangerous.

National Academy of Sciences. "Transgenic Plants and World Agriculture." National Academies Press. Available online. URL: http://www.nap.edu/html/transgenic/pdf/transgenic.pdf. Posted in July 2000. This report from an international working group of seven science academies, including five from developing nations, tells how agricultural biotechnology can relieve hunger and poverty in the developing world.

"National Opinion Poll on Labeling Genetically Engineered Foods." Center for Science in the Public Interest. Available online. URL: http://cspinet.org/reports/op_poll_labeling.html. Posted on May 16, 2001. Claims that

62 percent to 70 percent of Americans want genetically engineered foods labeled and that purchasing behavior would be affected by labeling of such foods.

"NIH Guidelines for Research Involving Recombinant DNA Molecules." National Institutes of Health Office of Biotechnology Activities. Available online. URL: http://www4.od.nih.gov/oba/rac/guidelines_02/NIH_Guidelines_Apr_02.htm. Last updated April 2002. Guidelines that all recombinant DNA research receiving federal funding must follow. Covers safety standards, roles and responsibilities, exemptions, requirements for protocols (plans for series of experiments), and more.

Persley, G. J. "New Genetics, Food and Agriculture: Scientific Discoveries—Societal Dilemmas." International Council for Science. Available online. URL: http://www.icsu.org/events/GMOs/index.html. Posted in June 2003. Analyzes findings of some 50 scientific reviews, published between 2001 and 2003, on the application of modern genetics to food, agriculture, and the environment. Conclusions are generally favorable to GM foods.

"Pharm and Industrial Crops: The Next Wave of Agricultural Biotechnology." Union of Concerned Scientists. Available online. URL: http://www.ucsusa.org/documents/PHARMcropsUCS403.pdf. Posted in March 2003. These crops promise new and cheaper drugs and other useful materials, but they also represent a greater potential threat to health and environment than present genetically modified crops and will require stricter government regulation.

"Public Attitudes to GM." Food Standards Agency. Available online. URL: http://www.foodstandards.gov.uk/multimedia/pdfs/gmfocusgroupreport.pdf. Posted on March 22, 2002. Reports findings from six focus groups convened by the agency, which advises the British government on food issues. The study was intended to determine the British public's feelings and beliefs about genetically modified foods.

"Regulating GMOs in Developing and Transition Countries." Food and Agriculture Organization of the United Nations (FAO). Available online. URL: http://www.fao.org/biotech/C9doc.htm. Posted on April 28, 2003. Background document provided for a conference on the subject beginning on that date. Discusses different kinds of genetically modified organisms, including crops, forest trees, animals, fish, and microorganisms; areas for regulation, including research and development, commercialization, commercial release, imports; existing international agreements and other factors affecting regulation; and balancing of costs and benefits.

Rutovitz, Jay, and Sue Mayer. "Genetically Modified and Cloned Animals: All in a Good Cause?" GeneWatch UK. Available online. URL: http://www.genewatch.org/GManimals/Reports/GManimalsRept.pdf.

Posted in April 2002. Focuses on ethical and animal welfare implications of the technology and concludes that most of its applications are not morally justified.

Sarageldin, Ismail, and G. J. Persley. "Promethean Science: Agricultural Biotechnology, the Environment, and the Poor." The World Bank Group. Available online. URL: http://www.worldbank.org/html/cgiar/publications/prometh/introd.pdf. Posted on May 2, 2000. Authors believe that biotechnology can help feed the world's poor in a sustainable, environmentally friendly way if it is accompanied by appropriate international policy. Recommendations including sequencing the genomes of important agricultural crops and placing this information in the public domain, identifying traits that affect productivity in marginal environments, conserving and characterizing existing genetic resources, and helping the developing world access technologies that will allow effective use of modified crops.

Smith, Nick. "Seeds of Opportunity: An Assessment of the Benefits, Safety, and Oversight of Plant Genomics and Agricultural Biotechnology." Lumen Foods. Available online. URL: http://www.soybean.com/docs/pressgmo.doc. Posted in April 2000. Claims that there is no significant difference between plants created by genetic engineering and similar ones created by traditional breeding and that U.S. regulation of bioengineered crops is adequate and should not be changed.

Taylor, Michael R., and Jody Tick. "Post-Market Oversight of Biotech Foods: Is the System Prepared?" Pew Initiative on Food and Biotechnology. Available online. URL: http://pewagbiotech.org/research/postmarket. Posted in April 2003. Claims that the ability of the FDA, EPA, and USDA to regulate genetically engineered crops after the crops have been approved for sale needs to be strengthened.

———. "The StarLink Case: Issues for the Future." Pew Initative on Food and Biotechnology. Available online. URL: http://pewagbiotech.org/resources/issuebriefs/starlink/starlink.pdf. Posted in October 2001. Analyzes the regulatory and public issues raised by this fall 2000 episode of contamination of the food supply by genetically engineered material not approved for human food.

"Three Years Later: Genetically Engineered Corn and the Monarch Butterfly Controversy." Pew Initiative on Food and Biotechnology. Available online. URL: http://pewagbiotech.org/resources/issuebriefs/monarch.pdf. Posted on May 30, 2002. Discusses the 1999 Cornell University study claiming that pollen from genetically engineered corn, which contains a built-in insecticide, killed monarch butterfly caterpillars; the subsequent outcry; and later research casting doubt on the initial report's conclusions.

"The Use of Genetically Modified Crops in Developing Countries." Nuffield Council on Bioethics. Available online. URL: http://www.nuffieldbioethics.org/filelibrary/pdf/gm_draft_paper.pdf. Posted in June 2003. Concludes that genetically modified crops can help small-scale farmers in developing countries and that there is a moral imperative to make such crops available easily and economically to those who want them.

Webster, Jocelyn, ed. "Foods from Genetically Improved Crops in Africa." San Diego Center for Molecular Agriculture. Available online. URL: http://www.sdcma.org/GIFoodsAfricaBrochure.pdf. Accessed on July 29, 2003. Provides background information on genetically modified crops and answers questions about their safety. Strongly supports their use in Africa and elsewhere.

"White Biotechnology: Gateway to a More Sustainable Future." EuropaBio. Available online. URL: http://www.europabio.org/upload/documents/wb_100403/Innenseiten_final_screen.pdf. Posted on April 10, 2003. Claims that crops genetically engineered to produce medicines and industrial products will have social, environmental, and economic benefits. Discusses policies for regulating such crops.

PATENTING LIFE

BOOKS

Albright, Matthew. *Profits Pending*. Monroe, Maine: Common Courage Press, 2002. Claims that biotechnology industry patents stifle new drug development, thereby indirectly killing people.

Andrews, Lori B., and Dorothy Nelkin. *The Body Bazaar: The Market for Human Tissue in the Biotechnology Age*. New York: Crown, 2001. Most people do not know it, but samples of their DNA and tissue are probably on file somewhere—and may be used for a variety of purposes without their permission. Andrews and Nelkin discuss the financial, social, and psychological issues raised by this situation, including questions of profit, property, privacy, and social control.

Cantuaria Marin, Patricia Lucia. *Providing Protection for Plant Genetic Resources: Patents, Sui Generis Systems, and Biopartnerships*. New York: Kluwer Law International, 2002. Analyzes international instruments and national laws dealing with patents, farmers' rights, and related subjects and shows how they affect developing countries rich in biodiversity and traditional knowledge, such as Brazil. Critiques several intercountry biopartnerships and suggests ways in which indigenous peoples and developing countries can achieve their fair share of benefits from partnerships.

Dutfield, Graham. *Intellectual Property Rights, Trade, and Biodiversity.* London: Earthscan Publications, 2002. Examines international agreements relating to intellectual property rights (IPRs) and makes recommendations for incorporating the requirements of the Convention on Biodiversity into the global regime for handling IPRs.

Magnus, David, Arthur L. Caplan, and Glenn McGee, eds. *Who Owns Life?* Loughton, Essex, U.K.: Prometheus Books, 2002. Anthology of essays discussing the legal, scientific, ethical, and economic issues involved in the patenting and "ownership" of genes, tissues, and organisms. Includes overview of intellectual property rights and history of legal developments in the field.

Office of Technology Assessment. *Ownership of Human Tissues and Cells: New Developments in Biotechnology.* Washington, D.C.: Books for Business, 2002. Analyzes the legal, economic, and ethical rights of the human sources of tissues and cells and of the physicians or researchers who obtain and develop these materials. Describes options for congressional action.

Santaniello, V., et al., eds. *Agriculture and Intellectual Property Rights: Economic, Institutional and Implementation Issues in Biotechnology.* Cambridge, Mass.: CABI Publishing, 2000. Policy makers and economists discuss issues in plant-breeding patents in biotechnology, the ownership of biological innovation, and associated intellectual property rights.

Shiva, Vandana. *Protect or Plunder? Understanding Intellectual Property Rights.* London: Zed Books, 2002. Activist Shiva discusses the 1980 U.S. Supreme Court decision to permit patenting of living things and the incorporation of intellectual property protection into the General Agreement on Tariffs and Trade (GATT), claiming that these exemplify trends that threaten the rights of indigenous people in developing countries.

Westerlund, Li. *Biotech Patents: Equivalency and Exclusions under European and U.S. Patent Law.* New York: Kluwer Law International, 2002. Compares laws in the United States, the European Community, Britain, and Germany regarding patent issues that affect biotechnology inventions, especially requirements for disclosure, infringement, and eligibility (including exclusions from patentability).

ARTICLES

Allen, Arthur. "Who Owns My Disease?" *Mother Jones*, vol. 26, November–December 2001, pp. 52ff. Critics say that in the scramble to patent human genes, companies have put profits ahead of patients' needs. Some families with congenital diseases are trying to rectify this by staking a claim on their own DNA.

Bobrow, Martin, and Sandy Thomas. "Patents in a Genetic Age." *Nature*, vol. 409, February 15, 2001, pp. 763–764. States that national and international patent systems need significant revision if they are not to hinder progress in biotechnology.

Brickley, Peg. "Payday for U.S. Plant Scientists." *The Scientist*, vol. 16, January 21, 2002, p. 22. Describes a December 2001 ruling in which the U.S. Supreme Court upheld the issuance of patents on genetically engineered plants.

Caulfield, Timothy A., and E. Richard Gold. "Whistling in the Wind." *Forum for Applied Research and Public Policy*, vol. 15, Spring 2000, pp. 75–76. Patenting of human genes, whether moral or not, is probably here to stay—but authors claim that the patent system can and should be changed to deal at least indirectly with moral issues.

Demaine, Linda J., and Aaron Xavier Fellmeth. "Reinventing the Double Helix: A Novel and Nonobvious Reconceptualization of the Biotechnology Patent." *Stanford Law Review*, vol. 55, November 2002, pp. 303–462. Argues that acceptance of patents on genes as desirable and legally proper rests on misinterpretation of three key concepts in patent law.

Doll, John J. "The Patenting of DNA." *Science*, vol. 280, May 1, 1998, pp. 689–690. Author, at the time the U.S. Patent and Trademark Office's Director of Biotechnology Examination, asserts that isolated and purified DNA sequences are "products of human ingenuity" and are therefore patentable.

Fleischer, Matt. "Patent Thyself." *American Lawyer*, vol. 23, June 2001, pp. 84–85. Describes how a group of families with a rare genetic disease is attempting to patent their tissues, which are useful to researchers, as a way of trying to make sure that tests or treatments developed from the tissues will remain affordable to patients.

Godrej, Dinjar. "Eight Things You Should Know About Patents on Life." *New Internationalist*, September 2002, pp. 9–12. Maintains that genes and living things belong to everyone and therefore should not be patented.

Hildyard, Nicholas, and Sarah Sexton. "No Patents on Life." *Forum for Applied Research and Public Policy*, vol. 15, Spring 2000, pp. 69ff. Maintains that patents on genes should not be granted because patenting inhibits research and locks up information that was often obtained with public money and therefore should be freely accessible.

Hughes, Sally Smith. "Making Dollars out of DNA." *Isis*, vol. 92, September 2001, pp. 541–576. Describes the first major biotechnology patent, granted in 1980 to the university employers of Stanley Cohen and Herbert Boyer for the basic process of recombinant DNA technology that they had developed. Shows how the patenting process was shaped by controversies of the time and in turn shaped the commercialization of molecular biology.

Hylton, Wil S. "Who Owns This Body?" *Esquire*, vol. 135, June 2001, pp. 102ff. Describes how human tissues and genes came to be patentable and what those patents may mean.

Marshall, Eliot. "Sharing the Glory, Not the Credit." *Science*, vol. 291, February 16, 2001, pp. 1189–1190. Arguments that occurred as public and private groups raced to produce a draft sequence of the human genome in early 2000 highlight philosophical disagreements about how genomic data should be shared.

Merson, John. "Bio-Prospecting or Bio-Piracy: Intellectual Property Rights and Biodiversity in a Colonial and Postcolonial Context." *Osiris*, 2000, pp. 282–297. Contrasts the granting of property rights to large companies for genetic and biological resources of developing countries, a "neo-colonialist" approach, with more equitable ways of handling these resources that are currently being explored.

Nenow, Lydia. "To Patent or Not to Patent: The European Union's New Biotech Directive." *Houston Journal of International Law*, vol. 23, Spring 2001, pp. 569ff. Argues that regulators of patent use, not patent issuers, should judge whether exploiting a given biotechnology invention is moral. Claims that present European patent laws, which allow issuers to reject patent requests if they consider the inventions immoral, will keep European countries from securing needed biotechnology capital.

Prakash, C. S. "Hungry for Biotech." *Technology Review*, vol. 103, July 2000, p. 32. Recommends that large biotechnology companies create improved versions of developing-world staple food crops such as rice and cassava and make them freely available to developing countries. They can then sell commercial crops to those countries with less public resistance.

Resnik, David B. "DNA Patenting and Human Dignity." *Journal of Law, Medicine, and Ethics*, vol. 29, Summer 2001, pp. 152ff. Concludes that patents on human DNA threaten but do not violate human dignity.

Shiva, Vandana. "North-South Conflicts in Intellectual Property Rights." *Synthesis/Regeneration*, Summer 2001, p. 28. Argues that multinational corporations and industrialized countries such as the United States abuse intellectual property protection and the rights of developing countries by patenting the fruits of indigenous knowledge. Cites a patent on the fungicidal use of India's neem tree, overturned by the European Patent Office in May 2000, as an example.

Shulman, Seth. "Toward Sharing the Genome." *Technology Review*, vol. 103, September 2000, pp. 60–61. Proposes five steps to balance commercial with public interests in patenting human genes, including declaring a moratorium on such patenting.

Wendt, Denise. "Canada's Supreme Court Says 'No' to Harvard Mouse." *BioPharm*, vol. 16, January 2003, p. 18. Describes the Supreme Court of

Canada's December 2002 decision not to allow a patent on a genetically engineered mouse or any other "higher" (multicellular) life form.

WEB DOCUMENTS

Commission of the European Communities. "Development and Implications of Patent Law in the Field of Biotechnology and Genetic Engineering." Europa—The European Union On-Line. Available online. URL: http://europa.eu.int/eur-lex/en/com/rpt/2002/com2002_0545en01.pdf. Posted on October 7, 2002. Examines the European Union's Directive 98/44 on patenting living things, highlighting compatibility with international agreements, patentability of plants and animals, patentability of elements isolated from the human body, and items excluded from patentability on ethical grounds. Concludes that the directive is an appropriate expression of society's concerns but that it needs to expand its scope to cover patenting human genes and stem cells.

"The Ethics of Patenting DNA." Nuffield Council on Bioethics. Available online. URL: http://www.nuffieldbioethics.org/filelibrary/pdf/theethicsofpatentingdna.pdf. Posted in July 2002. Summarizes round table discussions on this subject in June 2002. Argues that patents of DNA sequences should be the exception rather than the rule and that tests of inventiveness and usefulness should be more rigorously applied than they are at present.

"Genetic Inventions, Intellectual Property Rights and Licensing Practices." Organisation for Economic Co-operation and Development. Available online. URL: http://www.oecd.org/dataoecd/42/21/2491084.pdf. Posted December 22, 2002. Concludes that, despite some problems, the patent system basically achieves its aims in regard to biotechnological inventions. Offers recommendations for policy improvement.

"Higher Life Forms and the Patent Act." Canadian Biotechnology Advisory Committee. Available online. URL: http://cbac-cccb.ca/epic/internet/incbac-cccb.nsf/vwGeneratedInterE/ah00217e.html. Posted on February 24, 2003. Discusses the Canadian Supreme Court's ruling in December 2002 that "higher life forms," including plants and animals, are not patentable under present Canadian law.

HUGO Ethics Committee. "Statement on Benefit Sharing." Human Genome Organization. Available online. URL: http://www.gene.ucl.ac.uk/hugo/benefit.html. Posted on April 9, 2000. This statement by a respected international genetics organization recommends free access to genetic research information and urges companies that profit from such research to donate between 1 percent and 3 percent of their net profits to healthcare infrastructure or humanitarian causes.

Kevles, Daniel J. "A History of Patenting Life in the United States with Comparative Attention to Europe and Canada." Europa—the European Union On-Line. Available online. URL: http://europa.eu.int./comm/european_group_ethics/docs/study_kevles.pdf. Posted on January 12, 2002. Discusses the question of whether living things are patentable, the Plant Patent Act, the Chakrabarty case, plant and animal patents, ethical issues and policies in the United States as compared with Europe and Canada, and gene patenting.

"Public Agricultural Research: The Impact of IPRs on Biotechnology in Developing Countries." Food and Agriculture Organization of the United Nations (FAO). Available online. URL: http://www.fao.org/biotech/docs/torvergatareport.htm. Posted on June 27, 2002. Report of a workshop on the subject identifies constraints, needs, and opportunities within public sector research institutions; describes policy issues at institutional, national, and international levels; and identifies strategies to strengthen public sector biotechnology research for ensuring food security and alleviation of poverty.

Rural Advancement Foundation International (RAFI). "In Search of Higher Ground." The ETC Group. Available online. URL: http://www.etcgroup.org/documents/occ_higherground.pdf. Posted in September 2000. Claims that the patent system threatens public agricultural research and human rights and offers 28 alternative ways of protecting intellectual property.

HUMAN GENETICS

BOOKS

Alper, Joseph S., et al., eds. *The Double-Edged Helix: Social Implications of Genetics in a Diverse Society.* Baltimore, Md.: Johns Hopkins University Press, 2002. Discusses human gene research, mostly from perspectives outside science. Targets genetic determinism and warns against the possibility of genetic discrimination.

Avise, John C. *The Genetic Gods: Evolution and Belief in Human Affairs.* Cambridge, Mass.: Harvard University Press, 2001. Claims that genes control so much of human health, behavior, and thinking that they can be considered "gods," perhaps replacing those of traditional religions. Describes how evolution works at the genetic level and considers ethical issues raised by such subjects as the patenting of life, genetic screening and discrimination, gene therapy, and human cloning.

Beck, Stephen, and Alexander Olek, eds. *The Epigenome: Molecular Hide and Seek.* Hoboken, N.J.: Wiley, 2003. Describes cellular activity that mediates between genes and environment. This new area of study is expected to affect the findings of the Human Genome Project and their implications for society, industry, and health care.

Benjamin, Jonathan, Richard P. Ebstein, and Robert H. Belmaker, eds. *Molecular Genetics and the Human Personality.* Arlington, Va.: American Psychiatric Press, 2002. Considers the scientifically and ethically complex issue of the relationship between genes and personality, stressing that genes play a probabilistic rather than deterministic role in most cases.

Cherfas, Jeremy, ed. *Essential Science: The Human Genome.* London: DK Books, 2002. Explains the Human Genome Project and its implications for a nonscientific audience, using lively text and many graphics.

Coleman, William B., and Gregory J. Tsongalis, eds. *The Molecular Basis of Human Cancer.* Totowa, N.J.: Humana Press, 2001. Experts in basic and clinical science describe the genetics of several kinds of human cancer and consider future directions for research, including genetic counseling and gene therapy.

Davies, Kevin. *Cracking the Genome: Inside the Race to Unlock Human DNA.* Baltimore, Md.: Johns Hopkins University Press, 2002. Prominent science journalist describes the technologies and personalities involved in the Human Genome Project up to about 2000.

Dennis, Carina, and Richard Gallagher, eds. *The Human Genome.* New York: Palgrave Macmillan, 2002. Provides a concise overview of the science of the Human Genome Project, including its historical context, the public-private race to decipher the genome sequence, ways the genomic information may be used, and ethical issues these raise.

Ehrlich, Paul R. *Human Natures: Genes, Cultures, and the Human Prospect.* New York: Penguin, 2002. Overview for lay readers stresses that genes are greatly influenced by cultural and environmental factors in shaping human personality.

Finkler, Kaja. *Experiencing the New Genetics: Family and Kinship on the Medical Frontier.* Philadelphia: University of Pennsylvania Press, 2000. Criticizes the development of "genetic essentialism" and the medicalization of kinship, which has led to changes in ideas about the meaning and significance of the family in American society. Maintains that the genetic essentialist view can cause even healthy people to see themselves as sick and lead to abandonment of responsibility for environmental conditions that can also cause disease.

Lewontin, Richard C. *It Ain't Necessarily So: The Dream of the Human Genome and Other Illusions.* New York: New York Review of Books, 2001. Collection of reviews from the *New York Review of Books* written in the 1980s and 1990s by a geneticist famous for his outspoken opinions, plus replies and rebuttals from readers and updates on the relevant science.

The author is skeptical of the promised medical benefits of the Human Genome Project and critical of genetic determinism.

Marks, Jonathan. *What It Means to Be 98 Percent Chimpanzee: Apes, People, and Their Genes.* Berkeley: University of California Press, 2002. Criticizes both scientists who refuse to communicate with the public or consider social issues and nonscientists who subscribe to racism and other beliefs based on "folk heredity."

Peterson, Alan R., and Robin Bunton. *New Genetics and the New Public Health.* New York: Routledge, 2002. Explores the implications of recent genetic discoveries for medicine, public health, and health care.

Ridley, Matt. *Genome.* New York: HarperCollins, 2000. This book's 23 chapters, one for each of a human's pairs of chromosomes, provide a "whistle-stop tour" of the genome. The author uses the story of a gene from each chromosome to convey considerable information about genetic science and human development in an entertaining way.

———. *Nature via Nurture: Genes, Experience, and What Makes Us Human.* New York: HarperCollins, 2003. Shows how environment affects the way genes express themselves and how the two interact in a feedback loop to shape each human's uniqueness.

Rothman, Barbara Katz. *The Book of Life: A Personal and Ethical Guide to Race, Normality, and the Human Gene Study*, rev. ed. Boston: Beacon Press, 2001. Claims that genetics is, or can be, an ideology as well as a science—one with political and social implications, and one that is often overapplied. Warns that genetic determinism can lead to "new eugenics" and a way for individuals to avoid responsibility for their behavior and for society to ignore such problems as environmental pollution and poverty.

Sulston, John, and Georgina Ferry. *The Common Thread: A Story of Science, Politics, Ethics, and the Human Genome.* Washington, D.C.: Joseph Henry Press, 2002. The Nobel laureate who led the British team participating in the Human Genome Project provides a European insider's perspective on the project, including political as well as scientific maneuvering and disagreements about how genetic information should be shared.

Vogelstein, Bert, and Kenneth W. Kinzler, eds. *The Genetic Basis of Human Cancer.* New York: McGraw-Hill Professional, 2002. This collection of essays describes the complex picture that scientists now have of the genetic causation of different kinds of cancer, including possible treatments that may arise out of the research (among them, gene therapy).

Wade, Nicholas. *Life Script: How the Human Genome Discoveries Will Transform Medicine and Enhance Your Health.* New York: Touchstone Books, 2002. This book, primarily consisting of articles from the *New York Times*, describes the race to decipher the human genome and the likely medical and social effects the new information will have.

Wickelgren, Ingrid. *The Gene Masters: How a New Breed of Scientific Entrepreneurs Raced for the Biggest Prize in Biology.* New York: Times Books, 2002. This *Science* magazine correspondent describes the motivations and personalities of the major players in the Human Genome Project, focusing on the commercialization of the research process.

Zilinskas, Raymond A., and Peter J. Balint, eds. *The Human Genome Project and Minority Communities: Ethical, Social, and Political Dilemmas.* Westport, Conn.: Praeger, 2000. Eleven essays from a June 1991 conference examine the tension between minority communities' understandable skepticism about medical research in general and genetic research in particular and the benefits that new findings about human genetics might bring to groups statistically at high risk for certain gene-related health problems.

Zweiger, Gary. *Transducing the Genome: Information, Anarchy, and Revolution in the Biomedical Sciences.* New York: McGraw-Hill, 2001. Claims that the combination of information systems, business, and large-scale scientific projects, as exemplified by the Human Genome Project, will produce great benefits for everyone.

ARTICLES

Collins, Francis, and Victor A. McKusick. "Implications of the Human Genome Project for Medical Science." *Journal of the American Medical Association*, vol. 285, February 7, 2001, pp. 540ff. Describes the project, including its importance and work that remains to be done on it, and its implications for both experimental and mainstream medicine.

Coulter, Ian. "Genomic Medicine: The Sorcerer's New Broom?" *Western Journal of Medicine*, vol. 175, December 2001, pp. 424–426. Criticizes the new "genomic medicine" likely to grow out of findings of the Human Genome Project because it is reductionist and therefore ignores the whole person.

Dalrymple, Theodore. "'The Heart of My Mystery': Fear Not and Hope Not: The New Eugenics Will Leave Us Unchanged." *National Review*, vol. 53, July 9, 2001, p. 37. Stresses that genetic research and genetic engineering will never answer basic questions about human health and behavior.

Greenberg, Joel. "Covering the Human Genome Project." *The Quill*, vol. 89, April 2001, pp. 33–34. Background article for journalists describes scientific and ethical issues raised by the project.

"International Consortium Completes Human Genome Project." *Genomics & Genetics Weekly*, May 9, 2003, p. 32. The International Human Genome Sequencing Consortium announces the successful completion of the Human Genome Project more than two years ahead of schedule.

Lander, Eric S. "Genomics: Launching a Revolution in Medicine." *Journal of Law, Medicine, and Ethics*, vol. 28, Winter 2000, pp. S3–S4. Describes ways in which new knowledge of the human genome is beginning to transform the understanding and treatment of disease.

Nagel, Ronald L. "Molecule, Heal Thyself." *Natural History*, vol. 109, September 2000, pp. 30–34. Genetics is showing that the same disease can express itself differently in different people.

"The Proper Study of Mankind." *The Economist*, vol. 356, July 1, 2000, pp. 11ff. Claims that genetics and genomics have contributed considerably to the understanding of human history and prehistory and "human nature," including a reinterpretation of racial differences.

WEB DOCUMENTS

Baker, Catherine. "Your Genes, Your Choices." American Association for the Advancement of Science. Available online. URL: http://ehrweb.aaas.org/ehr/books/contents.html. Accessed on July 29, 2003. Aimed at young people or others with minimal science background. Discusses ethical issues raised by inherited diseases and gene therapy, genetic testing, forensic DNA databases, agricultural biotechnology, human gene enhancement, and human cloning.

"Blazing a Genetic Trail." Howard Hughes Medical Institute. Available online. URL: http://www.hhmi.org/genetictrail/. Updated spring 2002. Shows how researchers trace disease-causing genes through families and identify them, how genetic disorders are inherited, and what doctors are doing to prevent and treat them.

"Genetics and Human Behaviour." Nuffield Council on Bioethics. Available online. URL: http://www.nuffieldbioethics.org/publications/geneticsandhb/rep0000001098.asp. Posted on October 2, 2002. Considers ethical, social, and legal issues raised by research into behavioral genetics, including the possibility of changing or selecting unborn children based on genetic information related to behavioral traits and uses of behavior-related genetic information in the criminal justice system, education, employment, and insurance.

"Genomics and Its Impact on Science and Society: The Human Genome Project and Beyond." U.S. Department of Energy. Available online. URL: http://www.ornl.gov/TechResources/Human_Genome/publicat/primer/index.html. Posted in March 2003. This primer covers basic science, the Human Genome Project, what scientists have learned so far, legal and social implications, medical benefits, future scientific challenges, and more.

"Our Inheritance, Our Future: Realising the Potential of Genetics in the NHS." U.K. Department of Health. Available online. URL: http://www.

dh.gov.uk/assetRoot/04/01/92/39/04019239.pdf. Posted on June 24, 2003. Discusses the future role of genetics in Britain's National Health Service, including ways to realize the benefits of genetic research in health care and prevent inappropriate or unsafe use of genetic research and genetic information.

"Public Attitudes to Human Genetic Information." Human Genetics Commission. Available online. URL: http://www.hgc.gov.uk/business_publications_morigeneticattitudes.pdf. Posted in March 2001. A major survey of the British government's People's Panel finds broad support for the benefits promised by research on the human genome but concern about the regulation of such research.

DNA "FINGERPRINTING" AND DATABASES

BOOKS

Lee, Henry C., and Frank Tirnady. *Blood Evidence: How DNA Is Revolutionizing the Way We Solve Crimes.* Cambridge, Mass.: Perseus Publishing, 2003. Lee, chief criminalist for Connecticut for more than 20 years, shows how police collect DNA evidence from crime scenes and describes the history and methods of forensic DNA testing.

Scheck, Barry, Peter Neufeld, and Jim Dwyer. *Actual Innocence.* New York: Doubleday, 2000. Describes the Innocence Project, which attempts to prevent or reverse wrongful convictions through such means as DNA testing.

Spencer, Charlotte A. *Genetic Testimony: A Guide to Forensic DNA Profiling.* Upper Saddle River, N.J.: Prentice Hall, 2003. Explains DNA profiling for legal professionals or others with little background in science. Uses recent criminal cases to illustrate uses and potential misuses of DNA forensics.

ARTICLES

Amar, Akhil Reed. "A Safe Intrusion." *American Lawyer,* vol. 23, June 2001, p. 69. Supports creation of a universal database of limited DNA information as a way of avoiding conviction of the innocent and ensuring conviction of the guilty.

Beebee, Trevor. "Fingerprinting Wildlife." *Biological Sciences Review,* vol. 14, November 2001, pp. 26–30. Provides an example of the way ecologists use DNA fingerprinting, in this case to trace genetic variation and migration in a species of toad.

Burns, Thomas E. "An Overview of the National DNA Data Bank." *The Advocate*, vol. 59, May 2001, pp. 435–436. Describes the national forensic DNA databank that Canada established in June 2000.

Easterbrook, Gregg. "The Myth of Fingerprints—DNA and the End of Innocence." *The New Republic*, July 31, 2000, p. 20. Points out that death penalty foes praise DNA testing as a way to free the innocent, but such testing may ultimately support the death penalty by showing that most of those convicted of crimes really are guilty.

Guillen, Margarita, et al. "Ethical-Legal Problems of DNA Databases in Criminal Investigation." *Journal of Medical Ethics*, vol. 26, August 2000, pp. 266ff. Compares three types of forensic DNA databases: those with samples from all citizens, those with samples from people convicted of certain crimes, and those assembled on a case-by-case basis.

Hawkins, Dana. "Keeping Secrets." *U.S. News & World Report*, December 2, 2002, p. 58. Warns that DNA samples from ordinary people are accumulating in private databanks, where they may be misused.

Hollon, Tom. "Reforming Criminal Law, Exposing Junk Forensic Science." *The Scientist*, vol. 15, September 3, 2001, p. 12. Points out that state and federal DNA databases of criminals are helping to convict the guilty, free the innocent, and transform criminal law, but they would be even more useful if better funding removed backlogs.

Isenberg, Alice R. "Forensic Mitochondrial DNA Analysis: A Different Crime-Solving Tool." *FBI Law Enforcement Bulletin*, vol. 71, August 2002, pp. 16–22. Mitochondrial DNA analysis has both advantages and disadvantages compared to the more common analysis of nuclear DNA.

Jacot, Martine. "DNA in the Dock." *UNESCO Courier*, April 2000, pp. 37–38. Claims that creation of national DNA databases, even if they are limited to convicted criminals, is likely to lead to abuses of privacy and Fourth Amendment rights.

Kaye, D. H. "The Constitutionality of DNA Sampling on Arrest." *Cornell Journal of Law and Public Policy*, vol. 10, Summer 2001, pp. 55–109. Examines constitutional issues raised by taking, analyzing, and storing DNA samples and profiles from people arrested but not yet convicted of crimes. Concludes that a system for requiring arrested people to give DNA samples need not violate constitutional rights if it is carefully constructed to be minimally invasive and highly secure.

Kevles, Daniel J. "Ownership and Identity: The Drive to Manipulate DNA Has Changed the Economy and the Law." *The Scientist*, vol. 17, January 13, 2003, pp. 22–23. Points out that DNA identification and databases have many beneficial uses, as well as some potential abuses.

Kimmelman, Jonathan. "Just a Needle-Stick Away: DNA Testing Can Convict the Guilty; It Can Also Destroy the Privacy of Millions." *The Nation*, vol. 271, November 27, 2000, pp. 17–18. Describes ways in which law enforcement DNA databanks could threaten privacy rights.

———. "Risking Ethical Insolvency: A Survey of Trends in DNA Databanking." *Journal of Law, Medicine & Ethics*, vol. 28, Fall 2000, pp. 209ff. Reviews U.S. state statutes establishing forensic DNA databanks and compares them to laws in England and Canada. Identifies areas where statutes may inadequately protect rights to privacy, bodily integrity, and presumption of innocence and recommends ways to repair these weaknesses and reduce social risks of this powerful technology.

Linacre, Adrian. "The UK National DNA Database." *The Lancet*, vol. 361, May 31, 2003, p. 1841. Describes the March 2003 announcement by the British government that police would thereafter have the right to retain indefinitely all the DNA profiles they collect, whether or not the people from whom the profiles come are ever charged with a crime. Claims that the public seems confident that this extension of the national DNA database will not lead to wrongful suspicion or abuse of privacy.

Lyons, Donna, and Molly Burton. "Proof Positive." *State Legislatures*, vol. 27, June 2001, pp. 10ff. Reviews state laws related to DNA testing and the civil rights and other issues that such testing—or the lack of it—raises.

Palsson, Gisli, and Kristin E. Hardardottir. "For Whom the Cell Tolls: Debates About Biomedicine." *Current Anthropology*, vol. 43, April 2002, pp. 271–301. Places the debates about the genetic, genealogical, and medical databases being constructed for the population of Iceland in their domestic context, applying the moral perspective of economic anthropology.

Peterson, Rebecca Sasser. "DNA Databases: When Fear Goes Too Far." *American Criminal Law Review*, vol. 37, Summer 2000, pp. 1219ff. Discussion of forensic DNA databases raises the question of how far American society is prepared to go in allowing the government to impede liberty and privacy in the name of crime prevention. Claims that an all-citizen database threatens many freedoms and may be unconstitutional.

Raeturn, Paul. "Human Tissue: Handle with Care." *Business Week*, April 15, 2002, p. 75. Samples in tissue banks are often used in research without patients' consent, possibly compromising privacy and revealing sensitive genetic information.

WEB DOCUMENTS

"DNA Testing and the Death Penalty." American Civil Liberties Union. Available online. URL: http://www.aclu.org/DeathPenalty/DeathPenalty.cfm?ID=9315&c=65. Posted on June 26, 2002. Supports postconviction DNA testing as one way to ensure that innocent people are not sentenced to death.

Godard, Beatrice, et al. "Data Storage and DNA Banking for Biomedical Research." European Society of Human Genetics. Available online. URL: http://www.eshg.org/ESHG%20DNA%20banking%20bckgrnd.pdf. Posted in 2001. Covers informed consent, confidentiality, quality issues, ownership, and return of benefits. Recommendations are in a separate document (URL: http://www.eshg.org/ESHG%20DNA%20banking%20rec.pdf).

House of Lords. "Science and Technology Committee—Fourth Report." United Kingdom Parliament. Available online. URL: http://www.parliament.the-stationery-office.co.uk/pa/ld200001/ldselect/ldsctech/57/5701.htm. Posted on March 20, 2001. Discusses human genetic databases, including medical and forensic benefits, ethical issues related to privacy and consent, and personal use.

National Commission on the Future of DNA Evidence. "The Future of Forensic DNA Testing: Predictions of the Research and Development Working Group." National Criminal Justice Reference Service. Available online. URL: http://www.ncjrs.org/pdffiles1/nij/183697.pdf. Posted in November 2000. Discusses past and present techniques in forensic DNA analysis as well as the most likely technical advances in the coming decade and the probable impact of those advances.

GENETIC HEALTH TESTING AND DISCRIMINATION

BOOKS

Andrews, Lori B. *Future Perfect.* New York: Columbia University Press, 2002. Considers how the new understanding of the human genome will affect decisions regarding genetic testing and treatment. Examines three policy models (the medical model, the public health model, and the fundamental rights model) and concludes that the last, which requires informed consent, is best to apply to genetic testing.

Black, Edwin. *War Against the Weak: Eugenics and America's Campaign to Create a Master Race.* New York: Four Walls Eight Windows, 2003. New research on the history of eugenics ties Nazi atrocities firmly to the eugenics movement's earlier development in the United States. The author asserts that eugenics has been reborn in the second half of the 20th century as human genetics.

Khoury, Muin J., Wylie Burke, and Elizabeth J. Thomson, eds. *Genetics and Public Health in the 21st Century.* New York: Oxford University Press, 2000. Public health was the supposed justification for the eugenics (forced sterilization) laws of the early 20th century and other abuses of genetic information. This book's 31 contributions attempt to establish a framework for integration of recent advances in human genetics into public health practice that will avoid future social and ethical problems.

New York State Task Force on Life and the Law. *Genetic Testing and Screening in the Age of Genomic Medicine.* New York: New York State Task Force on Life and the Law, 2000. This report addresses ethical, legal, and social concerns surrounding predictive genetic testing and makes legislative, public policy, and practice recommendations regarding such issues as confidentiality, informed consent, insurers' use of genetic information, and use of clinical samples for genetic research.

Nordgren, Anders. *Responsible Genetics: The Moral Responsibility of Geneticists for the Consequences of Human Genetic Research.* New York: Kluwer Academic Publishers, 2003. Addresses moral questions regarding such subjects as genetic testing and counseling and eugenics from the seldom-considered perspective of the scientific community itself. Discusses different models of responsibility and makes proposals about the long-term goals of reprogenetic medicine.

ARTICLES

Baird, Patricia A. "Identification of Genetic Susceptibility to Common Diseases: The Case for Regulation." *Perspectives in Biology and Medicine,* vol. 45, Autumn 2002, pp. 516–528. Urges that the marketing of genetic tests to individuals be strongly regulated in order to avoid advertising that exaggerates the accuracy and benefits of such tests.

Bonn, Dorothy. "Genetic Testing and Insurance: Fears Unfounded?" *The Lancet,* vol. 355, April 29, 2000, p. 1526. Maintains that insurers seldom use genetic information and that, even when they do, the results are unlikely to have major effects on insurance rates.

Burke, W., L. E. Pinsky, and N. A. Press. "Categorizing Genetic Tests to Identify Their Ethical, Legal, and Social Implications." *American Journal of Medical Genetics,* vol. 106, issue 3, 2001, pp. 233–240. Makes recommendations to genetic counselors for handling different types of tests, depending on the accuracy of the test and whether treatment exists for the tested condition.

Calvo, Cheye. "From Laboratories to Legislatures." *State Legislatures,* vol. 27, September 2001, pp. 26–27. Considers the interactions among state laws, employers, and genetic testing, discussing such questions as when, if ever, genetic testing by employers should be permissible.

Ceniceros, Roberto. "Genetic Screening Faces Lawsuits." *Business Insurance,* vol. 35, February 19, 2001, p. 1. Describes the Equal Employment

Opportunity Commission's first lawsuit concerning genetic testing by an employer, which claimed violation of the Americans with Disabilities Act. The suit was filed against the Burlington Northern and Santa Fe Railroad Company, which tested certain employees for genetic predisposition to carpal tunnel syndrome after they filed injury claims related to this condition.

Frist, William. "Open Sesame." *Forum for Applied Research and Public Policy*, vol. 15, Spring 2000, pp. 44–45. New genetic information offers many useful possibilities, but it could also result in discrimination and other problems unless legislative steps are taken to prevent them.

Geetter, Jennifer S. "Coding for Change: The Power of the Human Genome to Transform the American Health Insurance System." *American Journal of Law and Medicine*, vol. 28, Spring 2002, pp. 1–76. Examines factors and players shaping current attitudes toward and legal regulation of genetic information, including the history of the eugenics movement. Concludes that genetic information is no different from other medical information and that risk-rated insurance should be abolished for all health traits, not just those with a genetic basis.

"Genetic Testing Results: Who Has a 'Right' to Know?" *Medical Ethics Advisor*, vol. 18, May 2002, pp. 49–52. Describes ethical dilemmas that can arise in genetic counseling, especially regarding effects of information on families, and suggests a "family covenant" as a way of handling them.

Gould, Stephen Jay. "Carrie Buck's Daughter." *Natural History*, July–August 2002, pp. 12–17. In this essay, first published in 1984, famed scientist-writer Gould offers evidence that the "three generations of imbeciles" involved in a eugenics lawsuit that reached the U.S. Supreme Court in 1927 were actually of normal intelligence.

Longman, Phillip J., and Shannon Brownlee. "The Genetic Surprise." *The Wilson Quarterly*, vol. 24, Autumn 2000, pp. 40ff. Maintains that either using or banning the use of genetic information will exacerbate weaknesses in the American health insurance and health care systems.

Martineau, Theresa M., and Caryn Lerman. "Genetic Risk and Behavioral Change." *British Medical Journal*, vol. 322, April 28, 2001, pp. 1056–1057. Discusses how testing for genetic risk of common diseases such as cancer may affect the health-related behavior of those tested.

McCabe, L. L., and E. R. McCabe. "Postgenomic Medicine: Presymptomatic Testing for Prediction and Prevention." *Clinical Perinatology*, vol. 28, issue 2, 2001, pp. 425–434. Infant screening programs are moving from biochemical tests for inherited conditions to direct testing of DNA or RNA and from tests for existing conditions to tests for increased risk of illness that may develop in the future. Authors maintain that these changes raise new issues of potential discrimination and privacy/confidentiality.

Meiser, Bettina, and Stewart Dunn. "Psychological Effect of Genetic Testing for Huntington's Disease: An Update of the Literature." *Western Journal of Medicine*, vol. 174, May 2001, pp. 336ff. Studies show that those found to carry the Huntington's gene (who are sure to develop the disease) and those found to be noncarriers differ significantly in short-term but not long-term psychological distress. Adjustment to results depended more on a person's psychological adjustment before testing than on what the results were.

Melzer, David, and Ron Zimmern. "Genetics and Medicalisation." *British Medical Journal*, vol. 324, April 13, 2002, pp. 863–864. Warns that doctors must be careful not to label someone as sick just because genetic screening shows that the person possesses a genetic variation that has been linked to increased risk of a disease; the person may never actually develop the condition.

Otchet, Amy. "Forsaking Genetic Secrets." *UNESCO Courier*, March 2001, pp. 29ff. Author describes her experience of being tested for the gene for Huntington's disease after her mother developed that devastating hereditary illness; she also discusses the possibility of discrimination in employment based on genetic testing.

Pagnatarro, Marisa Anne. "Genetic Discrimination and the Workplace: Employee's Right to Privacy v. Employer's Need to Know." *American Business Law Journal*, vol. 39, Fall 2001, pp. 139–186. Asserts that new national legislation is needed to protect workers from discrimination by employers and insurers on the basis of genetic information.

Parker, Michael, and Anneke Lucassen. "Concern for Families and Individuals in Clinical Genetics." *Journal of Medical Ethics*, vol. 29, April 2003, pp. 70–73. Discusses the ethical tensions between genetic counselors' responsibilities to individual patients and their responsibilities to other family members. Concludes that the issue must be resolved on a case-by-case basis.

Patterson, Annette, and Martha Satz. "Genetic Counseling and the Disabled: Feminism Examines the Stance of Those Who Stand at the Gate." *Hypatia*, vol. 17, Summer 2002, pp. 118–143. Examines the important role of genetic counseling in determining social attitudes toward people with disabling conditions, as well as the ways in which feminism may help counselors solve ethical dilemmas.

Pelias, M. K., and N. J. Markward. "Newborn Screening, Informed Consent, and Future Use of Archived Tissue Samples." *Genetic Testing*, vol. 5, issue 3, 2001, pp. 179–185. Most states mandate testing of infants for certain inherited diseases and do not require parental consent for the procedure, but researchers' later use of the resulting stored blood samples brings up ethical issues and should require some form of consent, authors maintain.

"Railroad Settles Suit over Testing to Predict Ergo Injuries." *Safety Compliance Letter,* June 15, 2002, p. 6. Describes mediated settlement in which the Burlington Northern and Santa Fe Railroad Company agreed to halt all genetic testing and pay $2.2 million to employees tested without their knowledge for a genetic predisposition to carpal tunnel syndrome.

Ridley, Matt. "Look Out, Prime Minister, That Napkin Could Be Dangerous!" *New Statesman,* May 13, 2002, pp. 29–31. Concludes that the likelihood of job discrimination and other abuses of genetic information has been exaggerated, but the danger does exist.

Ross, Lainie Friedman. "Ethical and Policy Issues in Genetic Testing." *Pancreatology,* vol. 1, 2001, pp. 576–580. Provides overview of six ethical issues raised by genetic testing and procurement of genetic information and recommends policies to address these issues.

Tauer, Jennifer Elle. "International Protection of Genetic Information." *Denver Journal of International Law and Policy,* Summer–Fall 2001, pp. 209–237. Claims that an international law is needed to protect people against genetic discrimination. UNESCO's declaration on the subject may be the best choice.

Tennant, Agnieszka. "Brave New Laws." *Books & Culture,* vol. 7, November–December 2001, pp. 12–13. Interview with Lori B. Andrews, an expert on the interactions of medicine, biotechnology, and the law, discusses ethical and legal aspects of genetic testing, ownership of body parts, and other biotechnology issues.

Therrell, B. L., Jr. "U.S. Newborn Screening Policy Dilemmas for the Twenty-first Century." *Molecular Genetic Metabolism,* vol. 74, issue 1–2, pp. 64–74. States have developed newborn screening programs for inherited diseases independently, with no national coordination or uniformity. A newborn screening task force identifies issues of concern about such programs and recommends a coordinated plan of action for revising them that involves public health programs, health care providers, and consumers.

Wheelwright, Jeff. "Testing Your Future." *Discover,* vol. 24, July 2003, pp. 35–40. Describes infant screening for inherited diseases, which every state now requires to some extent; at present the testing is biochemical, but it soon could become directly genetic, intensifying ethical issues and the risk of discrimination.

Wong, Josephine G., and Felice Lieh-Mak. "Genetic Discrimination and Mental Illness: A Case Report." *Journal of Medical Ethics,* vol. 27, December 2001, pp. 393–397. Describes a case in Hong Kong in which a court ruled that the civil service could not legally deny jobs in police and emergency services to people who had a family history of mental illness but showed no sign of disease themselves.

Yoon, P. W., et al. "Public Health Impact of Genetic Tests at the End of the Twentieth Century." *Genetic Medicine*, vol. 3, issue 6, 2001, pp. 405–410. Authors evaluated genetic tests in terms of their public health impact as measured by the number of people who potentially could be tested. They concluded that most genetic tests in current use have little effect on public health because they are for rare disorders, but as more genes are linked to common diseases, the public health impact of genetic testing is likely to increase.

Zachary, Mary-Kathryn. "Labor Law." *Supervision*, vol. 62, February 2001, pp. 22–23. Discusses laws and court cases relevant to employers' use of genetic testing.

WEB DOCUMENTS

Crosbie, Deborah. "Protection of Genetic Information: An International Comparison." Human Genetics Commission. Available online. URL: http://www.hgc.gov.uk/business_publications_international_regulations.pdf. Posted in September 2000. Reviews international laws and regulations concerning the protection of genetic information in a wide range of contexts, including DNA databanks, insurance, employment, forensics, research, and adoption.

"Ethical Issues with Genetic Testing in Pediatrics." American Academy of Pediatrics. Available online. URL: http://aapolicy.aapublications.org/cgi/content/full/pediatrics;107/6/1451.html. Posted on June 4, 2001. Reviews ethical considerations raised by use of genetic technology for newborn screening and for testing of children for carrier status or susceptibility to late-onset conditions. Recommends limited testing of children unless treatment is available for the conditions involved.

European Group on Ethics and Science in New Technologies. "Genetic Testing in the Workplace." Europa—The European Union On-Line. Available online. URL: http://europa.eu.int./comm/european_group_ethics/docs/opoce_en.pdf. Posted on March 6, 2000. Proceedings of a debate that discussed scientific, ethical, legal, and sociological aspects and heard the views of patients' associations, employees, and insurers.

Godard, Beatrice, et al. "Genetic Information and Testing in Insurance and Employment." European Society of Human Genetics. Available online. URL: http://www.eshg.org/ESHG%20insurance%20bckgrnd.pdf. Posted in 2002. Covers technical, social, and ethical issues. Recommendations are in a separate document (URL: http://www.eshg.org/ESHG%20insurance%20rec.pdf).

"Government Response to the Report from the House of Commons Science and Technology Committee: Genetics and Insurance." U.K. Department of Health. Available online. URL: http://www.dh.gov.uk/assetRoot/04/06/69/44/04066944.pdf. Posted on October 24, 2001. Association of British Insurers goes further than the government committee's request by agreeing on a five-year rather than a two-year moratorium on use of genetic test results for calculating insurance.

House of Commons. "Science and Technology—Fifth Report: Genetics and Insurance." Parliament Stationery Office. Available online. URL: http://www.parliament.the-stationery-office.co.uk/pa/cm200001/cmselect/cmsctech/174/17402.htm. Posted in March 2001. Makes a variety of recommendations, including a call for a voluntary two-year moratorium on insurers' use of genetic test results.

"Inside Information: Balancing Interests in the Use of Personal Genetic Data." Human Genetics Commission. Available online. URL: http://www.hgc.gov.uk/insideinformation/index.htm#report. Posted in May 2002. Report by commission advising the British government concludes that employers should not be allowed to require genetic tests, genetic discrimination should be legally banned, and testing someone's DNA or accessing a person's genetic information without the person's knowledge (except for legally permitted forensic uses) should be a criminal offense.

King, David. "The Persistence of Eugenics." Human Genetics Alert. Available online. URL: http://www.hgalert.org/topics/geneticDiscrimination/eugenics.htm. Accessed on July 31, 2003. Warns that a "new eugenics" may arise, either disguised as public health measures (as early 20th-century eugenics laws were in part) or in the form of free market demands.

"Newborn Screening: Characteristics of State Programs." General Accounting Office. Available online. URL: http://www.gao.gov/new.items/ d03449.pdf. Posted in March 2003. Reports that all states require screening for a minimum of four conditions. More than half have confidentiality provisions, but many have exceptions and may not be well enforced.

"Newborn Screening Tests." March of Dimes. Available online. URL: http://www.marchofdimes.com/professionals/681_1200.asp. Posted in May 2002. This factsheet answers parents' questions about screening of newborns for genetic diseases.

"The Regulatory Environment for Genetic Tests." Genetics and Public Policy Center. Available online. URL: http://www.dnapolicy.org/policy/genTests.jhtml. Posted April 2003. Covers the Clinical Laboratory Improvements Amendments of 1988, the FDA's role, protection of human subjects, regulations in the states, and the private sector.

Secretary's Advisory Committee on Genetic Testing. "Enhancing the Oversight of Genetic Tests." Office of Biotechnology, Activities, National Institutes of Health web site. Available online. URL: http://www4.od.

nih.gov/oba/sacgt/reports/oversight_report.pdf. Posted in July 2000. Discusses criteria for evaluation of genetic tests from both scientific and social points of view and recommends legislation and regulation to improve accuracy of tests and to protect testees from privacy intrusions and possible discrimination.

"Universal Declaration on the Human Genome and Human Rights." United Nations Educational, Scientific and Cultural Organization's Documentary Resources. Available online. URL: http://unesdoc.unesco.org/images/0010/001096/109687eb.pdf. Posted November 11, 1997. This November 1997 declaration asserts, among other things, that "the human genome in its natural state shall not give rise to financial gains" and affirms individuals' rights to have genetic tests or research performed on them only with consent, to be free of discrimination based on genetics, and to have their genetic information kept private.

"The Uses of Genetic and Other Predictive Medical Information in Insurance: The Handling of Rare Events." Genetic Interest Group. Available online. URL: http://www.gig.org.uk/docs/genetics_insurance.pdf. Posted on June 26, 2002. Proceedings of a workshop held on that date. Includes an insurer's perspective, clinical conclusions, family management, discussions, conclusions, and recommendations.

HUMAN GENE ALTERATION

BOOKS

Baldi, Pierre. *The Shattered Self: The End of Natural Evolution.* Cambridge, Mass.: MIT Press, 2001. Claims that technologies such as human gene alteration and cloning shatter the nature and concept of the human self as "a unique individual delimited by precise boundaries" but states that the self can continue to evolve after natural evolution has ended.

Beneke, Mark. *The Dream of Eternal Life.* Translated by Rachel Rubenstein. New York: Columbia University Press, 2002. Describes what recent research, especially in genetics, has revealed about the biological meanings of life, old age, and death, and also examines what these concepts mean on a cultural and psychological level. Considers the effects that extending human longevity or even achieving immortality through such techniques as cloning and cell transplants might have on humanity and the environment.

Bova, Ben. *Immortality: How Science Is Extending Your Life Span—and Changing the World.* New York: Avon, 2000. States that "the first immortals are already living among us. You might be one of them." Explains how new discoveries in biology may allow science to produce human immortality and considers the social implications of such a development.

Brodwin, Paul, ed. *Biotechnology and Culture: Bodies, Anxieties, Ethics.* Bloomington: Indiana University Press, 2001. Traces evolution of cultural debates over applications of biotechnology to the human body, such as cloning and prenatal genetic diagnosis, during the last hundred years in several locations.

Coors, Marilyn E. *The Matrix: Charting an Ethics of Inheritable Gene Alteration.* Lanham, Md.: Rowman & Littlefield, 2002. Examines ethics, religion, and genetic science to produce what the author calls the Inheritable Modification Matrix, which identifies virtues, values, and principles that differentiate beneficial from harmful uses of inheritable gene modification.

Davis, Dena S. *Genetic Dilemmas: Reproductive Technology, Parental Choices, and Children's Futures.* New York: Routledge, 2001. Considers ethical issues raised by disabled parents seeking assistance in conceiving a child who shares their disability, genetic testing of children that provides no immediate medical benefit to them, fetal sex selection, and human cloning.

Fukuyama, Francis. *Our Posthuman Future: Consequences of the Biotechnology Revolution.* London: Picador, 2003. Claims that limits on biotechnology are enforceable and desirable to prevent attempts to change human nature through genetic engineering or other means.

Hall, Stephen S. *Merchants of Immortality: Chasing the Dream of Human Life Extension.* Boston: Houghton Mifflin, 2003. Account of current research on life extension, stressing the personalities of scientists and entrepreneurs in the field and the effects of profit motives, politics, public distrust of science, and existing limitations on health care.

Harper, Joyce C., Joy D. A. Delhanty, and Alan H. Handyside, eds. *Preimplantation Genetic Diagnosis.* Hoboken, N.J.: Wiley, 2001. Scientists in fields relating to reproduction discuss different aspects of preimplantation genetic diagnosis of human embryos, including scientific background, specific procedures, and the future of ethics, regulation, and the approach as a whole.

Jedicke, Peter, ed. *Extreme Science: Transplanting Your Head and Other Feats of the Future.* New York: St. Martin's Press, 2001. Anthology of articles published in *Scientific American* between 1996 and 2000 covers subjects including gene therapy, cloning, and bionic bodies.

Kass, Leon R. *Life, Liberty, and the Defense of Dignity: The Challenge for Bioethics.* San Francisco: Encounter Books, 2002. Equates human dignity with "our awareness of need, limitation, and mortality" and fears that such future procedures as cloning and use of stem cells to regenerate failing bodies will alter the human race so much that this quality will no longer exist.

Lynn, Richard. *Eugenics: A Reassessment.* Westport, Conn.: Praeger, 2001. Argues that the late 20th century's condemnation of eugenics is not entirely justified: the eugenic objectives of eliminating genetic diseases, increasing intelligence, and reducing personality disorders are still desirable and will be achievable through biotechnology.

McKibben, Bill. *Enough: Staying Human in an Engineered Age.* New York: Times Books, 2003. Seeks the moral and spiritual boundary that the author calls the "enough point" in regard to future advances in genetic research, nanotechnology, and robotics. Claims that restraint is necessary to prevent, for example, children so genetically enhanced that they are no longer really human.

Maranto, Gina. *Quest for Perfection: The Drive to Breed Better Human Beings.* Lincoln, Neb.: iUniverse.com, 2000. Traces attempts to shape future generations from ancient societies' infanticide through Nazi "eugenic" mass murder to modern prospects of direct gene alteration. Suggests that science is going too far in altering human genetics.

Peterson, James C. *Genetic Turning Points: The Ethics of Human Genetic Intervention.* Grand Rapids, Mich.: William B. Eerdmans, 2001. Provides a Christian perspective on controversial issues related to human genetic testing and modification.

Roco, Mihail C., and William Sims Bainbridge, eds. *Converging Technologies for Improving Human Performance: Nanotechnology, Biotechnology, Information Technology and Cognitive Science.* New York: Kluwer Academic Publishers, 2003. Claims that the four sciences in the book's title have the potential to considerably enhance human performance and productivity.

Sloan, Philip R., ed. *Controlling Our Destinies: Historical, Philosophical, Ethical, and Theological Perspectives on the Human Genome Project.* Notre Dame, Ind.: University of Notre Dame Press, 2000. Analyzes the issues and values surrounding new knowledge about human genetics and the possibility of altering human genes.

Smith, George Patrick. *Bioethics and the Law.* Lanham, Md.: University Press of America, 2002. Considers ethical, medical, sociolegal, political, and philosophical aspects of issues raised by new knowledge about human genetics and the possibility of human gene alteration.

Stock, Gregory. *Redesigning Humans: Our Inevitable Genetic Future.* Boston: Houghton Mifflin, 2002. Claims that inheritable genetic enhancement of humans will become common whether governments attempt to ban it or not, but grants that the new technology will raise important ethical and scientific issues, including the potential for unexpected long-term effects of genetic changes.

———, and John H. Campbell, eds. *Engineering the Human Germline: An Exploration of the Science and Ethics of Altering the Genes We Pass to Our Children.* New York: Oxford University Press, 2000. Essays and discussions, mostly favoring the alteration of germ-line genes.

Wills, Christopher. *Children of Prometheus: The Accelerating Pace of Human Evolution.* Cambridge, Mass.: Perseus Publishing, 1999. Discusses how interactions between genes and environment have shaped human evolution so far and predicts an optimistic future for the species, partly guided by genetic and other scientific enhancement.

ARTICLES

Allen, Arthur. "How Far Would You Go for Healthy Kids?" *Redbook*, vol. 197, December 2001, pp. 102–107. Describes the controversy over using reproductive technology to avoid having a child with an inherited disease.

Ames, David A. "Eugenic Danger or Genetic Promise: A Revolution for the Millennium." *Cross Currents*, vol. 51, Fall 2001, pp. 293ff. Warns that the trend toward human gene alteration cannot be stopped and expresses hope that collaborative, interdisciplinary, interactive groups within communities of faith will grapple with the technology's ethical issues and potential dangers.

Anderson, W. French. "The Best of Times, the Worst of Times." *Science*, vol. 288, April 28, 2000, pp. 627–628. Stresses that gene therapy still holds great promise in spite of the fact that the death of a man in a gene therapy trial has cast a shadow over the field.

———. "A Cure That May Cost Us Ourselves." A pioneer in gene therapy looks forward to the possibility that such therapy will cure or prevent inherited diseases but states that the human germ line should never be altered.

Annas, George J., Lori B. Andrews, and Rosario M. Isasit. "Protecting the Endangered Human: Toward an International Treaty Prohibiting Cloning and Inheritable Alterations." *American Journal of Law and Medicine*, vol. 28, 2002, pp. 151–178. Summarizes recent international legal action in these areas and proposes an action plan and language for an international treaty banning human cloning and germ-line gene alterations.

Boyce, Nell. "The Cost of a Cure." *U.S. News & World Report*, January 27, 2003, p. 43. Development of leukemia in two French children who previously were among gene therapy's few successes suggests that some forms of the therapy substantially increase the risk of cancer.

———. "A Mother's Legacy." *U.S. News & World Report*, vol. 130, April 9, 2001, p. 52. A new assisted reproduction technique, mitochondrial replacement, may prevent certain rare inherited diseases but is controversial because it potentially alters germ-line genes.

Boyle, Robert J., and Julian Savulescu. "Ethics of Using Preimplantation Genetic Diagnosis to Select a Stem Cell Donor for an Existing Person." *British Medical Journal*, vol. 323, November 24, 2001, pp. 1240–1243. Claims that using preimplantation genetic diagnosis to choose a child to be a stem cell donor for an existing sibling is unlikely to result in harm and should be permitted.

Brownlee, Shannon. "Designer Babies." *Washington Monthly*, vol. 34, March 2002, pp. 25–31. Maintains that lack of regulation allows dubious techniques to be used in fertility clinics, including some that apparently alter germ-line genes.

Buckley, Rebecca H. "Gene Therapy for SCID—A Complication after Remarkable Progress." *The Lancet*, vol. 360, October 19, 2002, p. 1185. A French child treated successfully with gene therapy for severe combined immunodeficiency, an inherited disease, has developed leukemia. It remains to be seen whether this is a tragic accident or a new "black eye" for gene therapy.

Burtt, Shelley. "Which Babies?" *Tikkun*, vol. 16, January 2001, pp. 45ff. Claims that, rather than trying to eliminate children with imperfections before they are born, people need to rethink their willingness to live with unchosen obligations.

Christianson, Damaris. "Targeted Therapies: Will Gene Screens Usher in Personalized Medicine?" *Science News*, vol. 162, September 14, 2002, pp. 171–172. Explains that examination of patients' genes may help physicians predict the course of diseases and choose the best drugs and dosages to treat them with minimal side effects.

Cowley, G., and A. Underwood. "A Revolution in Medicine." *Newsweek*, vol. 135, April 10, 2000, pp. 58–59. Alteration of human genes may not become common for decades, but scientists are already using what they have learned about genetics to change the way a variety of diseases, from cancer to infections, are treated.

Deneen, Sally. "Designer People." *E*, vol. 12, Jan. 2001, pp. 26ff. Discusses ethical pros and cons of recent advances in human genetic engineering.

Dorsey, Michael. "The New Eugenics." *World Watch*, vol. 15, July–August 2002, pp. 21–24. States that a global public debate on the implications of technologies that alter human genes is greatly overdue. Recommends placing moratoria or bans on some of the more dangerous technologies in order to compel scientists and their supporters to consider more carefully the implications of their work.

d'Souza, Dinesh. "Staying Human: The Dangers of Techno-Utopia." *National Review*, vol. 53, Jan. 22, 2001. Replies to "techno-utopians" such as Lee Silver who favor human gene alteration that "in seeking to become gods, we are going to make monsters of ourselves."

———, and Ronald Bailey. "Our Biotech Future—An Exchange." *National Review*, vol. 53, March 5, 2001. Authors take opposing sides on the issue, with d'Souza attacking the technology, particularly in the form of alteration of human genes, and Bailey defending it.

Elfstrom, Gerard. "What Can Genetic Engineering Accomplish?" *Journal of the Alabama Academy of Science*, vol. 72, July 2001, pp. 190–196. Points out that recent discoveries in neuroscience suggest that attempts to produce particular human personality types by controlling genes or environment are likely to fail because so many complex interactions are involved.

Engelhardt, H. T., Jr. "Germline Engineering: The Moral Challenges." *American Journal of Medical Genetics*, vol. 108, issue 2, 2002, pp. 169–175. Argues that genetic engineering of the human germline is inevitable and that a clear distinction between medical use and enhancement is impossible. Claims that there is no convincing secular reason why the technology should be forbidden and, indeed, there may be some reasons why there is a moral obligation to employ it.

Foubister, Vida. "Intense Scrutiny Confronts Gene Therapy." *Medical World News*, vol. 43, February 28, 2000, pp. 1ff. In the wake of several deaths linked to gene therapy, critics claim that patients have been misled by their own enthusiasm and perhaps by researchers as well, but supporters question the link between the deaths and the therapy and say that, in any case, the deaths are relatively unimportant compared to the therapy's great promise.

Frankel, Mark S. "Inheritable Genetic Modification and a Brave New World: Did Huxley Have It Wrong?" *The Hastings Center Report*, vol. 33, March–April 2003, pp. 31–37. Claims that modification of germ-line genes is now proposed as a way to prevent disease, but market demand could easily lead to the technology's being used for enhancement, which may produce unpredictable effects on future generations.

Friedmann, Theodore. "Principles for Human Gene Therapy Studies." *Science*, vol. 287, March 24, 2000, pp. 2163ff. Analysis of human gene therapy trials in the wake of the death of Jesse Gelsinger concludes that tightening of regulations on such trials is highly desirable, but the technology has still achieved much..

Fukuyama, Francis. "In Defense of Nature." *World Watch*, vol. 15, July–August 2002, pp. 30–32. States that there are important reasons why cloning and human gene alteration should be of concern to everyone, especially those who wish to protect the natural environment.

Gillott, John. "Screening for Disability: A Eugenic Pursuit?" *Journal of Medical Ethics*, vol. 27, October 2001, p. 21. Claims that people who terminate a pregnancy because of fetal abnormalities are making judgments about their own and their children's lives, not about existing people with comparable abnormalities. Also states that motivation of the "new eugenics" is primarily medical, whereas that of early 20th century eugenics was primarily racial and social.

Gushee, David P. "A Matter of Life and Death." *Christianity Today*, vol. 45, October 1, 2001, pp. 34ff. Christian analysis of controversial biotechnology issues, including gene therapy, human cloning, and stem cell technology, supports the effort to use the technology to alleviate disease, as long as certain limits are respected.

Hudson, J. "What Kinds of People Should We Create?" *Journal of Applied Philosophy*, vol. 17, issue 2, 2000, pp. 131–143. Recommends that when human genetic engineering becomes practical, humans be altered to be more intelligent, motivated more by reason than by nonrational drives, and more resistant to pain.

Ikle, Fred C. "The Deconstruction of Death." *The National Interest*, Winter 2000, pp. 87ff. Claims that the greatest threats to the future are not human genetic alteration or cloning but the effects of considerably extending life and the creation of intelligent human-machine hybrids.

Launis, V. "Human Gene Therapy and the Slippery Slope Argument." *Medical Health Care Philosophy*, vol. 5, issue 2, 2002, pp. 169–179. Investigates validity of empirical, conceptual, and arbitrary result versions of the slippery slope argument and concludes that none provides a convincing reason for banning somatic cell gene therapy, although they do bring up important moral questions.

Levine, Judith. "What Human Genetic Modification Means for Women." *World Watch*, vol. 15, July–August 2002, pp. 26–29. Claims that feminists can support the right to abortion and still reject technologies of gene selection or alteration that could lead to a "new eugenics."

Lewis, Ricki. "Preimplantation Genetic Diagnosis: The Next Big Thing?" *The Scientist*, vol. 14, November 13, 2000, pp. 16–17. Couples can use preimplantation genetic diagnosis to select embryos free of genetic disease—or even to choose future babies' genders.

McKibben, Bill. "Design-a-Kid: Does Humanity Need an Upgrade?" *Christian Century*, vol. 120, May 17, 2003, pp. 22–28. Maintains that humans need to use their special ability to restrain themselves and avoid trying to improve the human genome.

Malpani, A., and D. Modi. "Preimplantation Sex Selection for Family Balancing in India." *Human Reproduction*, vol. 17, issue 1, 2002, pp. 11–12. Describes use of this technology in a society that strongly prefers boys and says that it need not produce gender disparity if it is limited to couples who already have at least one child and desire a child of the opposite sex.

Mooney, Pat. "Making Well People 'Better.'" *World Watch*, vol. 15, July–August 2002, pp. 13–17. Holds that pharmaceutical and biotechnology companies are more interested in creating drugs that enhance performance for the healthy and affluent than in helping the sick and poor.

Resnik, D. B., and P. J. Langer. "Human Germline Gene Therapy Reconsidered." *Human Gene Therapy*, vol. 12, issue 11, 2001, pp. 1449–1458.

Holds that germ-line gene modification, for any purpose, is not as morally problematic as some have claimed but that many ethical issues regarding its safety and efficacy still need to be addressed.

Savulescu, Julian. "Deaf Lesbians, 'Designer Disability,' and the Future of Medicine." *British Medical Journal*, vol. 325, October 5, 2002, pp. 771–773. Concludes that couples should have the right to apply genetic tests to future children and use the tests to choose which child they will have, even if their choice seems strange or even "wrong" to others.

———. "Procreative Benificence: Why We Should Select the Best Children." *Bioethics*, vol. 15, issue 5–6, 2001, pp. 413–426. Claims that this principle supports parents' use of preimplantation genetic diagnosis, not only to avoid disease in future children, but to choose any traits that are likely to give the child what they feel will be the best possible life.

Savulescu, Julian, and E. Dahl. "Sex Selection and Preimplantation Genetic Diagnosis: A Response to the Ethics Committee of the American Society of Reproductive Medicine." *Human Reproduction*, vol. 15, issue 9, 2000, pp. 1879–1880. The committee concluded that use of preimplantation genetic diagnosis (PGD) for sex selection or any other nonmedical purpose should be discouraged, but authors argue that the technology should be available for this use, at least within privately funded health care.

SerVaas, Cory, and Patrick Perry. "Coronary Gene Therapy: Restoring Hope and Lives." *Saturday Evening Post*, vol. 272, May 2000, pp. 38ff. Describes a promising form of gene therapy that inserts a growth-promoting gene into coronary arteries of people suffering from blood vessel blockage, which produces new blood vessels in the area and reduces painful angina.

Singer, Peter A., and Abdallah S. Daar. "Harnessing Genomics and Biotechnology to Improve Global Health Equity." *Science*, vol. 294, October 5, 2001, pp. 87–88. Proposes a five-point strategy for harnessing genome-related biotechnology to improve health in developing countries.

Swift, E. M., and Don Yeager. "Unnatural Selection." *Sports Illustrated*, vol. 94, May 14, 2001, pp. 86–92. Describes controversial near-future use of genetic enhancement in professional sports.

Thompson, Larry. "Human Gene Therapy: Harsh Lessons, High Hopes." *FDA Consumer*, vol. 34, September 2000, pp. 19ff. Recounts the often painful history of gene therapy.

Trefil, James. "Brave New World." *Smithsonian*, December 2001, pp. 38–46. Overview of advances and ethical concerns regarding assisted reproduction, gene therapy, genetic engineering, cloning, and stem cell research.

Tudge, Colin. "The Future of Humanity." *New Statesman*, vol. 131, April 8, 2002, pp. 25–27. Concludes that genetic engineering will not be able to change humanity to any great degree and that this is fortunate.

Watson, James, et al. "Future Visions: How Will Genetics Change Our Lives?" *Time*, vol. 161, February 17, 2003, pp. 60–61. Pioneers in today's genetic revolution predict its effects during the next 50 years, particularly in medicine.

Wheelwright, Jeff. "Betting on Designer Genes." *Smithsonian*, January 2001, pp. 29–36. Describes the promise of gene therapy and explains why fulfilling it is proving harder than anyone had thought.

———. "Body, Cure Thyself." *Discover*, vol. 23, March 2002, pp. 62–68. During the decade after the first attempt at human gene therapy, the field has seen both improvements and setbacks.

Wolfson, Adam, and Ronald Bailey. "Does Genetic Engineering Endanger Human Freedom?" *American Enterprise*, vol. 12, October 2001, pp. 40ff. Claims that genetic engineering of humans is not intrinsically harmful and can be supported by the principles of autonomy and benificence. Warns against government control of the technology.

WEB DOCUMENTS

Deech, Ruth, and Helena Kennedy. "Outcome of the Public Consultation on Preimplantation Genetic Diagnosis." Human Genetics Commission. Available online. URL: http://www.hgc.gov.uk/business_publications_pgdoutcome.pdf. Posted in November 2001. This report of a public consultation carried out by the Human Genetics Commission and the Human Fertilisation and Embryology Authority, both agencies of the British government, shows public support for parents' use of preimplantation genetic diagnosis (PGD) to avoid having children with serious genetic disorders but also highlights concern about the technology's wider implications.

Frankel, Mark S., and Audrey R. Chapman. "Human Inheritable Genetic Modifications: Assessing Scientific, Ethical, Religious, and Policy Issues." American Association for the Advancement of Science. Available online. URL: http://www.aaas.org/spp/sfrl/projects/germline/report.pdf. Posted in September 2000. Recommends creation of an independent body to oversee research into human germ-line gene modification as well as extensive public discussion of the scientific and moral issues involved.

The Genetics and Public Policy Center with Princeton Survey Research Associates. "Public Awareness and Attitudes About Reproductive Genetic Technology." Genetics and Public Policy Center. Available online. URL: http://dnapolicy-content.labvelocity.com/pdfs/4/27374.pdf. Posted on December 9, 2002. Reports on a survey that explored the knowledge and attitudes of 1,211 respondents about reproductive cloning, genetic testing, and genetic modification and preferences about government regulation.

"The Regulatory Environment for Human Gene Transfer." Genetics and Public Policy Center. Available online. URL: http://www.dnapolicy.org/policy/humanGeneTransfer.jhtml. Posted in April 2003. Discusses federal and state regulations relating to human gene transfer.

HUMAN CLONING AND EMBRYONIC STEM CELL RESEARCH

BOOKS

Bonnicksen, Andrea. *Crafting a Cloning Policy: From Dolly to Stem Cells.* Washington, D.C.: Georgetown University Press, 2002. Examines political response to advances in cloning and stem cell research, discusses interest groups' attempts to shape policy, and identifies four types of cloning policy.

Brannigan, Michael C., ed. *Ethical Issues in Human Cloning: Cross-Disciplinary Perspectives.* New York: Seven Bridges Press, 2000. This book is divided into sections on scientific, religious, philosophical, and legal issues.

Cibelli, Jose, et al., eds. *Principles of Cloning.* San Diego, Calif.: Academic Press, 2002. Articles by principal investigators of experiments in cloning animals describe how cloning works and consider applications of the technique in basic biology, agriculture, biotechnology, and medicine.

Dudley, William, ed. *At Issue: The Ethics of Human Cloning.* San Diego, Calif.: Greenhaven Press, 2001. Includes articles by experts in ethics, religion, law, politics, science, and technology.

Fritz, Sandy, ed. *Understanding Cloning.* New York: Warner Books, 2002. Fifteen articles from *Scientific American* provide a brief overview of the science, applications, and ethics of cloning technology.

Klotzko, Arlene J. *A Clone of Your Own? The Science and Ethics of Cloning.* New York: Oxford University Press, 2002. Concludes that human reproductive cloning is presently morally indefensible because of physical problems revealed in animal cloning experiments.

———, ed. *The Cloning Sourcebook.* New York: Oxford University Press, 2001. Twenty-seven essays on the science and context of cloning and the ethical and policy issues it raises.

Kristol, William, and Eric Cohen, eds. *The Future Is Now: America Confronts the New Genetics.* Lanham, Md.: Rowman & Littlefield, 2002. Chronicles the debate over embryonic stem cell research, human cloning, and other aspects of recent research on human genetics. Most authors are conservative and are not scientists, but they are anything but unified in their opinions, particularly on stem cell research.

Lauritzen, Paul, ed. *Cloning and the Future of Human Embryo Research.* New York: Oxford University Press, 2001. Shows that the debate about cloning is inextricably tied to those on reproductive technology and research on embryos.

MacKinnon, Barbara, ed. *Human Cloning: Science, Ethics, and Public Policy.* Champaign: University of Illinois Press reissue, 2002. Papers from a 1998 conference on scientific, ethical, and policy issues related to human cloning.

McGee, Glenn, ed. *The Human Cloning Debate,* 3d ed. Albany, Calif.: Berkeley Hills Books, 2002. New edition includes material on cloning to produce stem cells as well as animal and human reproductive cloning.

National Research Council. *Stem Cells and the Future of Regenerative Medicine.* Washington, D.C.: National Academies Press, 2002. Workshop held in June 2001 concludes that research on both adult and embryonic stem cells, including research cloning, should proceed.

Peat, F. David. *Scientific and Medical Aspects of Human Reproductive Cloning.* Washington, D.C.: National Academies Press, 2003. National Academy of Sciences panel reviews the scientific and medical, but not the religious, societal, or ethical, issues raised by human cloning. It concludes that reproductive cloning should be prohibited because it is dangerous and likely to fail, but cloning of embryos for the purpose of harvesting stem cells for research should be permitted.

Prentice, David A., and Michael A. Palladino. *Stem Cells and Cloning.* San Francisco, Calif.: Benjamin Cummings, 2002. Defines stem cells and their potential applications, distinguishing between embryonic and adult stem cells, and discusses the scientific, ethical, and political issues raised by their use.

Ruse, Michael, and Christopher A. Pynes, eds. *The Stem Cell Controversy: Debating the Issues.* Loughton, Essex, U.K.: Prometheus Books, 2003. Presents a wide variety of perspectives on the medical and ethical issues raised by research on embryonic stem cells, including the harvesting of cells from aborted fetuses.

ARTICLES

Bayliss, F. "Human Cloning: Three Mistakes and an Alternative." *Journal of Medical Philosophy,* vol. 27, issue 3, 2002, pp. 319–337. Among the mistakes the author cites is thinking of cloning as strictly a reproductive technology; she holds that it is also an enhancement technology.

Bottum, J. "The Horror." *Public Interest,* Winter 2003, pp. 54–59. Claims that embryonic cloning for research is even more immoral than reproductive cloning because it makes human embryos "plastic playthings for the human will."

Brock, D. W. "Human Cloning and Our Sense of Self." *Science*, vol. 296, 2002, issue 5566, pp. 314–316. Replies to lines of argument stating that human reproductive cloning would undermine the sense of individuality, human worth, and freedom or autonomy.

Brown, Barry L. "Human Cloning and Genetic Engineering: The Case for Proceeding Cautiously." *Albany Law Review*, vol. 65, Spring 2002, pp. 649–677. Discusses the scientific history of cloning, the risks and benefits of human cloning, a constitutional framework for determining whether to support or deny individual instances of cloning, and the present legislative response to the prospect of reproductive and research cloning. Suggests that an absolute ban on human cloning is neither ethically desirable nor workable and that a supportive but firm regulatory environment is better.

Bush, George W. "Address to the Nation on Stem Cell Research from Crawford, Texas." *Weekly Compilation of Presidential Documents*, vol. 37, August 13, 2001, p. 1149. President Bush explains his decision to allow federal funding for research on existing embryonic stem cell lines but not for creation or use of any new lines.

Cibelli, Jose B., et al. "The First Human Cloned Embryo." *Scientific American*, vol. 286, January 2002, pp. 44–51. Advanced Cell Technology's announcement that it had cloned human embryos, which developed to the six-cell stage before dying.

"Cloning: A Report and Its Lessons." *National Review*, vol. 54, August 12, 2002, p. 16. Claims that the arguments of the members of the President's Council on Bioethics who favor cloning of embryos for stem cell research show the weakness of the ethical position that destructive research on embryos is permissible.

Cregan, Kate. "Stem Cells—Cleaving at the Root?" *Arena Magazine*, April–May 2002, pp. 13–14. Claims that research on embryonic stem cells will probably benefit only the wealthy and that "a ball of cells, a collection of embryonic tissue, does matter."

Dunn, Kyla. "Cloning Trevor." *Atlantic Monthly*, vol. 289, June 2002, pp. 31–46. Inside report of research at the biotechnology company Advanced Cell Technology, which resulted in cloning of the first human embryos but did not produce embryos that developed far enough to allow harvesting of stem cells.

Durrell, Justine. "Can the Law Handle Human Cloning?" *Trial*, vol. 38, October 2002, pp. 24–30. Surveys existing laws and court decisions that cover or might relate to human research or reproductive cloning. Concludes that new causes of action are likely to result once human cloning of either variety begins to occur.

Feldbaum, Carl. "Some History Should Be Repeated." *Science*, vol. 295, February 8, 2002, p. 975. The president of the Biotechnology Industry

Organization (BIO) explains why the BIO supports research cloning but rejects reproductive cloning.

Fischer, Joannie. "The First Clone." *U.S. News & World Report*, December 3, 2001, pp. 50ff. Provides background on Advanced Cell Technology's claim that it has produced the first cloned human embryos, including profiles of the people involved.

Groopman, Jerome. "Holding Cell—Why the Cloning Decision Was Wrong." *The New Republic*, August 5, 2002, p. 14. Maintains that arguments for placing a four-year moratorium on cloning of human embryos for stem cell research are weak and unnecessarily delay medical advances.

Hansen, J. E. "Embryonic Stem Cell Production through Therapeutic Cloning Has Fewer Ethical Problems Than Stem Cell Harvest from Surplus IVF Embryos." *Journal of Medical Ethics*, vol. 28, issue 2, 2002, pp. 86–88. Holds that, based on ethical considerations alone, therapeutic cloning should be encouraged.

Johnson, Alissa. "Attack of the Clones." *State Legislatures*, vol. 29, April 2003, pp. 30–33. Review of state laws concerning human cloning.

Jonietz, Erika. "Cloning, Stem Cells, and Medicine's Future." *Technology Review*, vol. 106, June 2003, pp. 70–71. In this interview the president and chief executive officer (CEO) of the biotechnology company Geron explains why he thinks that research on human embryos, especially on embryonic stem cells, is essential for medical advancement.

Kass, Leon R. "Preventing a Brave New World." *Human Life Review*, vol. 27, Summer 2001, pp. 14ff. Holds that the hotly debated issue of human cloning may provide a "golden opportunity" for people to think carefully about ethical issues raised by genetic manipulation and attempt to regain ethical control over the technology.

———. "A Reply." *Public Interest*, Winter 2003, pp. 58–63. Kass replies to comments about the conclusions regarding human cloning and embryonic stem cell research that the President's Council on Bioethics reached in its July 2002 report.

Katayama, A. "Human Reproductive Cloning and Related Techniques: An Overview of the Legal Environment and Practitioner Attitudes." *Journal of Assisted Reproductive Genetics*, vol. 18, issue 8, pp. 442–450. In a survey, most assisted reproductive technology professionals said they would be both willing and able to clone a human child if it were legal to do so.

Keenan, Faith. "Cloning: Huckster or Hero?" *Business Week*, July 1, 2002, p. 86. Describes Michael West, head of the biotech company Advanced Cell Technology. West frequently has been accused of exaggerating his company's achievements, including its research cloning of human embryos, but he insists that his motives are ethical.

Klotzko, Arlene Judith. "Take Therapeutic Cloning Forward." *The Scientist*, vol. 11, July 22, 2002, pp. 16–17. Claims that therapeutic (research) cloning should proceed, but it should be carefully regulated, as in Senator Dianne Feinstein's proposed bill, which author says is similar to regulations in Britain.

Krauthammer, Charles. "Crossing Lines—A Secular Argument Against Research Cloning." *The New Republic*, April 29, 2002, pp. 20–21. Opposes research (stem cell) cloning because it would create human embryos for the sole purpose of destroying them, making them in effect mere manufactured "things."

Miller, Jon D. "Breaking News or Broken News: A Brief History of the 'First Cloned Human Embryo' Story." *Nieman Reports*, vol. 56, Fall 2002, pp. 18–21. Discusses the reporting of Advanced Cell Technology's November 2001 announcement that the company had created the first cloned human embryos. Cites several lessons to be learned from these events, including the need to apply caution and scientific expertise to claims before publishing stories about them.

Peters, Ted, and Gaymon Bennet. "Theological Support of Stem Cell Research." *The Scientist*, vol. 15, September 3, 2001, p. 4. Concludes that there is an active ethical and theological mandate to support such research because of its potential use in medicine.

"Protein Arrangements May Make Primate Cloning Impossible." *Chemistry and Industry*, April 21, 2003, p. 7. Scientists attempting to clone rhesus monkeys have found that the nuclear process usually employed in animal cloning and proposed for human cloning appears to damage primate eggs in a way that does not happen with eggs of other mammals.

Rifkin, Jeremy. "Why I Oppose Human Cloning." *Tikkun*, vol. 17, July–August 2002, pp. 23–26. The author says he objects to research (stem cell) cloning of human embryos because it could create an exploitative market for women's eggs and increase treatment of living things as commodities, and also because adult stem cells should work just as well in medical treatments as embryonic cells.

Schaub, Diana. "Slavery Plus Abortion." *Public Interest*, Winter 2003, pp. 41–46. Strongly opposes human cloning, including embryonic cloning for stem cell research, calling the latter "the willful destruction of nascent human life."

Shannon, Thomas A. "The Rush to Clone: It's Unethical to Apply This Unproven Research to Humans." *America*, vol. 185, September 10, 2001, pp. 8ff. Claims that cloning humans is unethical because the technology has not been proven safe in animals.

"Stem Cells and Xenotransplantation: Ethics, Patents, and Politics, an Industry Roundtable." *BioPharm*, vol. 15, June 2002, pp. 19ff. Transcript of a discussion by experts in the biotechnology industry, including representatives of a biotechnology company and a global law firm.

Stock, Gregory, and Francis Fukuyama. "The Clone Wars: A *Reason* Online Debate." *Reason,* vol. 34, June 2002, pp. 34–41. The authors discuss cloning and other forms of human gene alteration ("designer babies"), with Stock supporting such actions because of their potential to prevent or cure disease and Fukuyama opposing them because of their potential to create social inequality.

Vaughan, Christsopher, and Kevin Cool. "Cell Division." *Stanford,* May–June 2003, pp. 43–49. Scientists working at Stanford University's new institute for cancer and stem cell research find themselves at the center of the debate on the ethics of cloning embryos to produce possible medical treatments.

Weinberg, Robert A. "Of Clones and Clowns." *Atlantic Monthly,* vol. 289, June 2002, pp. 54–58. Maintains that decisions about human cloning, both reproductive and research (stem cell), are being made on the basis of media hype rather than properly evaluated science.

"The Wrong Road: Cloning and Stem Cells." *The Economist,* August 4, 2001, p. 14. Concludes that the United States needs an independent regulatory body, like Britain's Human Fertilisation and Embryology Authority, to oversee genetic science, especially applications to humans such as human cloning.

Young, Cathy. "Monkeying Around with the Self." *Reason,* vol. 32, April 2001, pp. 23ff. Creation of the first genetically engineered primate and predictions that a cloned human being may be imminent raise ethical issues about human cloning that both sides of the debate tend to oversimplify.

WEB DOCUMENTS

Alexander, Brian. "(You)^2." *Wired* Magazine. Available online. URL: http://www.wired.com/wired/archive/9.02/projectx_pr.html. Posted in February 2001. Like it or not, a cloned human child is likely to come into existence soon, this author says.

Chapman, Audrey R., Mark S. Frankel, and Michele S. Garfinkel. "Stem Cell Research and Applications: Monitoring the Frontiers of Biomedical Research." American Association for the Advancement of Science. Available online. URL: http://www.aaas.org/spp/sfrl/projects/stem/report.pdf. Posted in November 1999. Focuses on potential benefits of stem cell research, especially on embryonic stem cells, but recognizes that such research presents ethical and regulatory problems. Recommends use of excess embryos from in vitro fertilization, with informed consent of donor couples, rather than creation of new embryos.

Department of Health and Human Services, National Institutes of Health. "Stem Cells: Scientific Progress and Future Research Directions." Stem Cell Information, National Institutes of Health. Available online. URL: http://stemcells.nih.gov/stemcell/scireport.asp. Updated on June 4, 2003. Describes significant progress in research on both adult and embryonic stem cells and considers medical uses to which the cells might be put.

European Group on Ethics and Science in New Technologies. "Ethical Aspects of Human Stem Cell Research and Use." Europa—The European Union On-Line. Available online. URL: http://europa.eu.int/comm/european_group_ethics/docs/dp15rev.pdf. Revised version posted in September 2002. Considers the scientific and medical aspects, ethical issues and recommendations by ethics committees, legal aspects, viewpoints of several major religions, and regulatory frameworks.

"FAQs About Cloning, Stem Cell Research, and ASRM's Position." American Society of Reproductive Medicine. Available online. URL: http://www.asrm.org/Media/misc_announcements/cloning/SCNT.html. Accessed on July 29, 2003. Explains why the organization opposes human reproductive cloning but supports embryonic stem cell research.

House of Lords Stem Cell Research Committee. "Stem Cell Research—Report" United Kingdom Parliament. Available online. URL: http://www.parliament.the-stationery-office.co.uk/pa/ld200102/ldselect/ldstem/83/8301.htm. Posted on February 13, 2002. Stresses the importance of research using embryonic stem cells and supports the government Human Fertilisation and Embryology Authority's decision to allow therapeutic cloning, but prefers the use of embryos left over from in vitro fertilization to creation of new embryos. Discusses national and international regulatory aspects.

"Human Cloning and Human Dignity: An Ethical Inquiry." President's Council on Bioethics. Available online. URL: http://www.bioethics.gov/reports/cloningreport/index.html. Posted in July 2002. This major report by 17-member advisory panel to the U.S. government thoughtfully examines the ethical issues brought up by two forms of human cloning. All members opposed human reproductive cloning for safety and many other reasons, but members were divided about research or therapeutic cloning, which would create cloned human embryos for the purpose of harvesting stem cells from them (and thus destroying them) at a very early stage. Seven members of the council wanted research cloning to proceed under regulation, but the other 10 voted to place a four-year moratorium on the process so public dialogue could continue.

Pence, Gregory E. "Ten Myths About Human Cloning." Reproductive Cloning Network. Available online. URL: http://www.reproductive-cloning.net/open/myths.html. Posted in 2001. Explains why cloned humans would not simply be "xeroxes" or in any other way less human or individual than nonclones.

"The Regulatory Environment for Human Cloning." Genetics and Public Policy Center. Available online. URL: http://www.dnapolicy.org/policy/humanCloning.jhtml. Posted in April 2003. Covers federal laws restricting cloning, Food and Drug Administration regulation of human reproductive cloning, state regulations related to cloning, and organizational policies and recommendations.

"Stem Cell Basics." Stem Cell Information, National Institutes of Health. Available online. URL: http://stemcells.nih.gov/infoCenter/stemCellBasics.asp. Updated September 2002. Describes unique properties of all stem cells, compares adult and embryonic stem cells, and surveys potential uses of stem cells and the obstacles that must be overcome before these uses can be achieved.

"Stem Cell Research: Second Update." The Royal Society. Available online. URL: http://www.royalsoc.ac.uk/files/statfiles/document-148.pdf. Posted in June 2001. Updates Royal Society reports "Therapeutic Cloning" and "Stem Cell Research and Therapeutic Cloning: An Update."

"Stem Cell Research and Therapeutic Cloning: An Update." The Royal Society. Available online. URL: http://www.royalsoc.ac.uk/files/statfiles/document-123.pdf. Posted in November 2000. Supports proposed legislative controls that would prevent reproductive cloning but permit research (therapeutic) cloning.

"Stem Cell Therapy: The Ethical Issues." Nuffield Council on Bioethics. Available online. URL: http://www.nuffieldbioethics.org/filelibrary/doc/stem_cell_therapy2.doc. Posted in April 2000. This paper, based on a roundtable meeting held in September 1999, examines key ethical issues raised by stem cell therapy and makes recommendations.

Stephens, Patrick. "Cloning: Towards a New Conception of Humanity." Reproductive Cloning Network. Available online. URL: http://www.reproductivecloning.net/open/objectivist.html. Posted in January 2001. Answers criticisms of the ethics of human reproductive cloning by Leon Kass and others.

"Therapeutic Cloning: Submission by the Royal Society to the Chief Medical Officer's Expert Group." The Royal Society. Available online. URL: http://www.royalsoc.ac.uk/files/statfiles/document-104.pdf. Posted in March 2000. Focuses on stem cells because of their application to a wide variety of potential medical treatments.

CHAPTER 8

ORGANIZATIONS AND AGENCIES

There are many organizations devoted to biotechnology/genetic engineering and various aspects of genetics. The following entries include professional groups, industry organizations, advocacy groups, and government agencies, both in the United States and abroad. In keeping with the widespread use of the Internet and e-mail, the web site address (URL) and e-mail address are given first when available, followed by the phone number and postal address.

AfricaBio
URL: http://www.africabio.com/index.shtml
E-mail: africabio@mweb.co.za
Phone: (+27) 12 667 2689
Promotes safe, ethical, and responsible research, development, and application of biotechnology and its products. Stresses safety and benefits of the technology but also supports labelling of genetically modified foodstuffs.

African Agricultural Technology Foundation (AATF)
URL: http://www.aftechfound.org
E-mail: aatfinformation@ilri.exch.cgiar.org
Phone: 254-020-630743
African Agricultural Technology Foundation c/o ILRI
P.O. Box 30709
Nairobi 00100, Kenya
Africa-based and managed public-private partnership linking large agricultural biotechnology companies such as Monsanto with small-holder farmers in Africa to bring the benefits of biotechnology to that continent and improve plant varieties grown by the country's small farmers.

AgBioWorld Foundation
URL: http://www.agbioworld.org
Phone: (334) 444-7884
P.O. Box 85
Tuskegee Institute, AL 36087-0085
Headed by C. S. Prakash of Tuskegee University. Highly in favor of agricultural biotechnology

but does not accept direct contributions from biotechnology companies or trade associations.

Agricultural Biotechnology Council (abc)
URL: http://www.abcinformation.org
E-mail: Enquiries@abcinformation.org
Phone: (+44 0) 207 395-8944
P.O. Box 38 589
London SW1A 1WE
United Kingdom
Established in 2002 by the British biotechnology trade industry, including large international companies, to provide information and education that address public concerns about the technology.

Agricultural Groups Concerned About Resources and the Environment (AGCare)
URL: http://www.agcare.org
E-mail: agcare@agcare.org
Phone: (519) 837-1326
192 Nicklin Road
Guelph, Ontario N1H 7L5
Canada
Coalition of groups representing Ontario's farmers. It provides science and research-based information and policy initiatives on subjects including crop biotechnology, which it supports.

Agriculture and Environment Biotechnology Commission (AEBC)
URL: http://www.aebc.gov.uk
E-mail: aebc@dti.gsi.gov.uk
Phone: (+440) 207 215 6508
Bay 479
1 Victoria Street
London SW1H 0ET
United Kingdom
Provides the British government with independent, strategic advice on developments in biotechnology and their implications for agriculture and the environment, including ethical and social as well as scientific issues. Its activities include meetings, reports, and public debate.

Alliance for Bio-Integrity
URL: http://www.bio-integrity.org/
E-mail: info@biointegrity.org
Phone: (641) 472-5554
12040 Pearl Lane #2
Fairfield, IA 52556
Believes that genetically engineered foods present "unprecedented dangers to the environment and human health" and should be labeled and carefully regulated.

American Civil Liberties Union (ACLU)
URL: http://www.aclu.org
E-mail: aclu@aclu.org
Phone: (212) 549-2500
125 Broad Street
18th Floor
New York, NY 10004-2400
One area of concern for this nonprofit, nonpartisan public interest organization devoted to protecting American civil liberties is genetic privacy. It opposes employment and insurance discrimination based on genetic testing as well as the establishment of national DNA databases. It offers both print and online

resources, including books, pamphlets, newsletters, and Web links.

American College of Medical Genetics (ACMG)
URL: http://www.acmg.net
E-mail: acmg@acmg.net
Phone: (301) 634-7127
9650 Rockville Pike
Bethesda, MD 20814-3998
Professional organization for scientists and health care professionals specializing in medical genetics. Among other things, provides guidelines for genetic health testing and lobbies for effective and fair health policies and legislation.

American Corn Growers Association (ACGA)
URL: http://www.acga.org
E-mail: acga@acga.org
Phone: (202) 835-0330
P.O. Box 18157
Washington, DC 20036
Progressive commodity association, representing thousands of corn producers in 28 states. Questions whether the benefits of genetically modified corn outweigh its economic liability and other risks to farmers.

American Farm Bureau
URL: http://www.fb.com
Phone: (202) 406-3660
600 Maryland Avenue, SW
Suite 800
Washington, DC 20024
Implements policies and provides programs that improve the financial well-being and quality of life for farmers and ranchers. Supports agricultural biotechnology and opposes labeling of genetically modified foods.

American Genetic Association
URL: http://www.theaga.org/overview.html
E-mail: agajoh@mail.ncifcrf.gov
Phone: (301) 695-9292
P.O. Box 257
Buckeystown, MD 21717-0257
Promotes basic and applied research on the genetics of plants and animals. Publishes *Journal of Heredity.*

American Society for Reproductive Medicine (ASRM)
URL: http://www.asrm.org
E-mail: asrm@asrm.org
Phone: (205) 978-5000
1209 Montgomery Highway
Birmingham, AL 35216-2809
Professional organization devoted to advancing knowledge and expertise in reproductive medicine and biology. Publishes educational and other materials, including some that deal with ethical considerations.

American Society of Gene Therapy (ASGT)
URL: http://www.asgt.org
E-mail: info@asgt.org
Phone: (414) 278-1341
611 E. Wells Street
Milwaukee, WI 53202
Fosters education, exchange of information, and research on gene therapy.

American Society of Human Genetics
URL: http://genetics.faseb.org/genetics/ashg/ashgmenu.htm
E-mail: society@ashg.org
Phone: (301) 634-7300
9650 Rockville Pike
Bethesda, MD 20814-3998
Professional society of researchers, physicians, genetic counselors, and others interested in human genetics and related social issues. Publishes monthly *American Journal of Human Genetics.*

American Society of Law, Medicine, and Ethics (ASLME)
URL: http://www.aslme.org
E-mail: info@aslme.org
Phone: (617) 262-4990
765 Commonwealth Avenue
Suite 1634
Boston, MA 02215
Members include attorneys, physicians, health care administrators, and others interested in relationship between law, medicine, and ethics. Publishes *American Journal of Law* and *Journal of Law, Medicine, and Ethics,* both quarterly.

American Soybean Association (ASA)
URL: http://www.soygrowers.com
Phone: (800) 688-7692
12125 Woodcrest Executive Drive
Suite 100
St. Louis, MO 63141-5009
Trade association that aims to improve profitability for U.S. soybean farmers and to supply soy products to the world. Supports raising of genetically engineered soybeans.

BioIndustry Association (BIA)
URL: http://www.bioindustry.org/
E-mail: admin@bioindustry.org
Phone: (44) 20 7565 7190
14/15 Belgrave Square
London SW1X 8PS
United Kingdom
British biotechnology trade association that represents the industry and its needs to audiences ranging from patient groups to regional, national, and pan-European governments.

BIOTECanada
URL: http://www.biotech.ca/EN
E-mail: info@biotech.ca
Phone: (613) 230-5585
130 Albert Street
Suite 420
Ottawa, Ontario K1P 5G4
Canada
National biotechnology industry lobbying group devoted to promoting a better understanding of biotechnology and its benefits to Canadians.

Biotechnology Industry Organization (BIO)
URL: http://www.bio.org
Phone: (202) 962-9200
1225 I Street, NW
Suite 400
Washington, DC 20005
Chief trade and lobbying organization for the biotechnology industry, including academic institutions as well as commercial companies.

British Society for Human Genetics (BSHG)
URL: http://www.bshg.org.uk
E-mail: bshg@bshg.org.uk
Phone: (440) 121 627-2634
Clinical Genetics Unit
Birmingham Women's Hospital
Birmingham B15 2TG
United Kingdom
Professional association for British scientists and health professionals involved in clinical and laboratory research on human genetics. Its Public Policy Committee considers the ethical and social effects of human genetic research and technology. It publishes a quarterly newsletter and issues statements on aspects of human genetics.

The Campaign to Label Genetically Engineered Foods
URL: http://www.thecampaign.org/index.php
E-mail: label@thecampaign.org
Phone: (425) 771-4049
P.O. Box 55699
Seattle, WA 98155
National grassroots consumer campaign for legislation that will require labeling of genetically engineered food in the United States. As of 2003, the organization focuses on city and county administrations and state legislatures rather than on the federal government.

Canadian Biotechnology Advisory Committee
URL: http://cbac-cccb.ca
E-mail: info@cbac-cccb.ca
Phone: (866) 748-2222
240 Sparks Street
West Tower
Fifth Floor
Suite 561E
Ottawa, Ontario K1A 0H5
Canada
Provides expert advice to the federal government on ethical, social, regulatory, economic, scientific, environmental, and health aspects of biotechnology.

Center for Food Safety (CFS)
URL: http://www.centerforfoodsafety.org
E-mail: office@centerforfoodsafety.org
Phone: (202) 547-9359
660 Pennsylvania Avenue, SE
Suite 302
Washington, DC 20003
Nonprofit membership organization concerned about impacts of the current industrial food production system on human health, animal welfare, and the environment. Conducts legal and grassroots campaigns to, among other things, protect consumers from what it sees as the hazards of genetically engineered foods.

Center for Genetics and Society
URL: http://www.genetics-and-society.org/index.asp
E-mail: info@genetics-and-society.org
Phone: (510) 625-0819
436 14th Street
Suite 1302
Oakland, CA 94612
Nonprofit information and public affairs organization working to en-

courage responsible uses and effective governance of new human genetic and reproductive technologies. Supports medical applications of these technologies but opposes applications that objectify and commodify human life and threaten to divide human society or alter the processes of the natural world.

Center for Science in the Public Interest (CSPI)
URL: http://www.cspinet.org
E-mail: cspi@cspinet.org
Phone: (202) 332-9110
1875 Connecticut Avenue, NW
Suite 300
Washington, DC 20009
Nutrition advocacy organization; publishes *Nutrition Action Healthletter.* Its Biotechnology Project is aimed at accurately identifying the risks and benefits of agricultural biotechnology and strengthening FDA regulation of genetically engineered food crops, for instance by making company submission of safety data mandatory rather than voluntary.

Centre for Applied Bioethics
URL: http://www.nottingham.ac.uk/bioethics/
E-mail: ben.mepham@nottingham.ac.uk
Phone: (440 115) 951-6303
School of Biosciences
University of Nottingham
Sutton Bonington Campus
Loughborough
Leicester LE12 5RD
United Kingdom
Concerned with appropriate application of biotechnology to food production, industry, and medical uses of farm animals.

Clonaid
URL: http://www.clonaid.com
E-mail: pr@clonaid.com
Phone: (+1 702) 497-9186
Clonaid, which calls itself the first human cloning company, was founded by the Raelians, a group that believes that extraterrestrials used genetic engineering to create all life on Earth. The company and its director, Brigitte Boisselier, attracted considerable media attention when they announced the birth of a cloned baby, Eve, in December 2002, but they never proved their assertion.

Clone Rights United Front
URL: http://www.clonerights.com
E-mail: r.wicker@verizon.net
Phone: (212) 255-1439
506 Hudson Street
New York, NY 10014
Claims that the right to be cloned is part of the right to control reproduction. Opposes bans on human cloning.

Coalition for the Advancement of Medical Research (CAMR)
URL: http://www.stemcellfunding.org/fastaction
E-mail: CAMResearch@yahoo.com
Phone: (202) 833-0355
2021 K Street, NW
Suite 305
Washington, DC 20006

The coalition includes patient organizations, universities, scientific societies, foundations, and individuals with life-threatening disorders. It supports advances in regenerative medicine, including stem cell research and somatic cell nuclear transfer, and works to ensure that research cloning remains legal and receives federal funding. It opposes reproductive cloning.

Codex Alimentarius Commission
URL: http://www.codexalimentarius.net
E-mail: Codex@fao.org
Phone: (+39) 065-7051
Viale delle Terme di Caracalla
00100 Rome
Italy
Established in 1963 by the United Nations Food and Agriculture Organization (FAO) and World Health Organization (WHO) to develop food standards, guidelines, and codes of practice that protect consumer health and ensure fair trade practices in the food trade. Its work includes safety assessments of genetically modified foods.

Consultative Group on International Agricultural Research (CGIAR)
URL: http://www.cgiar.org
E-mail: cgiar@cgiar.org
Phone: (202) 473-8951
CGIAR Secretariat
The World Bank
MSN G6-601
1818 H Street, NW
Washington, DC 20433
Research organization, with both public and private members, whose mission is to contribute to food security and poverty eradication in developing countries through research, partnerships, capacity building, and policy support, promoting sustainable agricultural development based on the environmentally sound management of natural resources. It supports a system of 16 Future Harvest Centers, working in more than 100 countries. Some of its projects apply genetic engineering to crops raised by farmers in the developing world.

Consumer Federation of America (CFA)
URL: http://www.consumerfed.org
Phone: (202) 387-6121
1424 16th Street, NW
Suite 604
Washington, DC 20036
Represents more than 285 consumer organizations. Gathers, analyzes, and disseminates information to the public, legislators, and regulators and advocates on behalf of consumers, especially the least affluent. Holds conferences and issues reports, books, brochures, news releases, and a newsletter. Approves of biotechnology but wants it carefully regulated.

Consumer Project on Technology (CPTech)
URL: http://www.cptech.org
Phone: (202) 387-8030

P.O. Box 19367
Washington, DC 20036
Group started by Ralph Nader. Its web site includes discussion of biotechnology and gene patents related to health care technologies, which it generally opposes or would like to see limited.

Council for Agricultural Science and Technology (CAST)
URL: http://www.cast-science.org/cast/src/cast_top.htm
E-mail: cast@cast-science.org
Phone: (515) 292-2125
4420 West Lincoln Way
Ames, IA 50014-3447
Collects, interprets, and communicates science-based information on food, fiber, agricultural, natural resource, and related societal and environmental issues to the public, the private sector, the media, policy makers, regulators, and legislators. Includes information on agricultural biotechnology.

Council for Biotechnology Information
URL: http://www.whybiotech.com
Phone: (202) 467-6565
1225 I Street, NW
Suite 400
Washington, DC 20043-0380
Established by the leading biotechnology companies and trade associations to communicate science-based information about the benefits and safety of agricultural and food biotechnology.

Council for Biotechnology Policy (CBP)
URL: http://www.pfm.org/BiotechTemplate.cfm
Phone: (703) 478-0100
Bioethics organization sponsored by the Wilberforce Forum, a division of Prison Fellowship Ministries. Works to provide information, present a Christian perspective, and shape policy on bioethical issues affecting human dignity, such as cloning and stem cell research.

Council for Responsible Genetics (CRG)
URL: http://www.gene-watch.org
E-mail: crg@essential.org
Phone: (617) 868-0870
5 Upland Road Suite 3
Cambridge, MA 02140
National nonprofit organization of scientists, public health professionals, and others that works to see that biotechnology develops safely and in the public interest. Its concerns include genetic discrimination, patenting of life-forms, food safety, and environmental quality. Publishes a bimonthly newsletter, GeneWatch, and educational materials.

Council of Canadians
URL: http://www.canadians.org
E-mail: inquiries@canadians.org
Phone: (800) 387-7177
502-151 Slater Street
Ottawa, Ontario K1P 5H3
Canada
Citizen watchdog organization. Regards claims of biotechnology benefits as "too good to be true."

CropLife America
URL: http://www.croplifeamerica.org
Phone: (202) 296-1585
1156 15th Street, NW
Suite 400
Washington, DC 20005
Trade association that promotes innovation and the environmentally sound manufacture, distribution, and use of crop protection and production technologies. Stresses the benefits of genetically engineered crops.

Cultural Survival
URL: http://www.cs.org
E-mail: culturalsurvival@cs.org
Phone: (617) 441-5400
215 Prospect Street
Cambridge, MA 02139
Defends human rights and cultural autonomy of indigenous peoples and oppressed ethnic minorities, including protection from exploitation by multinational biotechnology companies. Publishes *Cultural Survival Quarterly*, *Cultural Survival Curriculum Resources*, and two other journals.

Department of Agriculture (USDA)
Animal and Plant Health Inspection Service (APHIS)
URL: http://www.aphis.usda.gov/
E-mail: APHIS.web@aphis.usda.gov
Phone: (301) 734-7607
4700 River Road
Riverdale, MD 20737-1236
U.S. government agency that regulates any organisms, including genetically altered organisms, that are or might be plant pests. Offers publications on genetically engineered plants and related subjects.

Department for Environment, Food and Rural Affairs (DEFRA)
URL: http://www.defra.gov.uk
E-mail: helpline@defra.gsi.gov.uk
Phone: (44) 0 20 7238 6951
Information Resource Centre
Lower Ground Floor
Ergon House
c/o Nobel House
17 Smith Square
London SW1P 3JR
United Kingdom
British governmental agency that took over the functions of the Ministry of Agriculture, Fisheries, and Food. Among other things, DEFRA is responsible for the control of the deliberate release of genetically modified organisms (GMOs), and for national, EU, and international policy on the environmental safety of GMOs.

Ecological Farming Association (EFA)
URL: http://www.eco-farm.org
E-mail: info@eco-farm.org
Phone: (831) 763-2111
406 Main Street
Suite 313
Watsonville, CA 95076
Nonprofit educational organization that promotes ecologically sound agriculture. Has joined other organic farming and environmental organizations in protesting against

genetically modified organisms and the consolidation of farming production into the hands of large multinational corporations.

Environmental Defense
URL: http://www.environmentaldefense.org/home.cfm
E-mail: members@environmentaldefense.org
Phone: (212) 505-2100
257 Park Avenue South
New York, NY 10010
National nonprofit organization that links science, economics, and law to create innovative, equitable, and cost-effective solutions to urgent environmental problems. Usually opposes genetically modified crops as threats to the environment and, possibly, human health.

Environmental Protection Agency (EPA)
URL: http://www.epa.gov/
E-mail: public-access@epa.gov
Phone: (202) 227-0167
Ariel Rios Building
1200 Pennsylvania Avenue, NW
Washington, DC 20460
U.S. government agency that regulates pesticides, including biopesticides, and sets limits for the amounts of such substances that can remain on or in food; also regulates new chemicals, which are considered to include some genetically engineered organisms.

The ETC Group
URL: http://www.etcgroup.org
Phone: (204) 453-5259
478 River Avenue
Suite 200
Winnipeg, Manitoba R3L OC8
Canada
Formerly Rural Advancement Foundation International (RAFI). Dedicated to conservation and sustainable advancement of cultural and ecological diversity and human rights. Provides research and analysis of technical information, including information about plant genetic resources, biotechnology, and biodiversity, and works to develop strategic options for applying those technologies at the global and regional (international) levels in ways that will create cooperative and sustainable self-reliance within disadvantaged societies.

European Association for Bioindustries (EuropaBio)
URL: http://www.europabio.org/pages/index.asp
E-mail : mail@europabio.org
Phone: (+32 2) 735-0313
Avenue de l'Armée 6
1040 Brussels
Belgium
International trade association representing smaller associations and companies of all sizes working in a variety of biotechnology fields. Works to promote innovative and dynamic yet responsible biotechnology-based industry in Europe and to advocate free and open markets for the technology.

European Community of Consumer Cooperatives (Euro Coop)
URL: http://www.eurocoop.org/home/en/default.asp
E-mail: info@eurocoop.coop
Phone: (32) 2-285-0070
Rue Archimède, 17
B - 1000 Brussels
Belgium
Consumer organization that has developed close relations with the European Commission, the European Parliament, and the European Economic and Social Committee and represents consumer interests in committees that advise these groups. Stresses that most European consumers do not want to eat genetically modified foods and expresses concern about keeping organic and conventional farming free of GM contamination.

European Federation of Biotechnology (EFB)
URL: http://www.efbweb.org
E-mail: info@efbweb.org
Phone: (31) 15 212 7800
Oude Delft 60 NL-2611 CO Delft
The Netherlands
Works to increase public and government understanding of biotechnology. Publications include a newsletter and briefing papers on patenting life, biotechnology in foods and drinks, the application of human genetic research, and environmental technology.

European Group on Ethics and Science in New Technologies (EGE)
URL: http://europa.eu.int/comm/european_group_ethics/index_en.htm
E-mail: GOPA-ETHICS-GROUP@cec.eu.int
Phone: (32 2) 299-9696
European Commission
Brey 10/128 B-1049 Brussels
Belgium
Independent multidisciplinary body that advises the European Commission on ethical aspects of science and new technologies. It has provided opinions on various subjects, including human tissue banking, human embryo research, personal health data in the information society, and human stem cell research.

European Molecular Biology Organization (EMBO)
URL: http://www.embo.org
E-mail: embo@embo.org
Phone: (49) 6221 88910
Postfach 1022.40
D-69012 Heidelberg
Germany
Established in 1964, EMBO promotes molecular biology in Europe and neighboring countries. Publishes the *EMBO Journal* and EMBO Reports. Its Science and Society Committee communicates with the nonscientific community about the effects and benefits of molecular biology. It sponsors research and training through the European Molecular Biology Laboratory and its outstations.

European Society of Human Genetics (ESHG)
URL: http://www.eshg.org
E-mail: eshg@eshg.org

Phone: (43) 1-405-138322
c/o Vienna Medical Academy
Alser Strasse 4
1090 Vienna, Austria
Arranges meetings and other scientific gatherings in the field of human genetics and publishes *The European Journal of Human Genetics.*

Extropy Institute (ExI)
URL: http://www.extropy.org
E-mail: natasha@extropy.org
Phone: (512) 263-2749
10709 Pointe View Drive
Austin, TX 78738
Promotes new technology and human advancement in all forms, including genetic enhancement.

Federal Bureau of Investigation, Combined DNA Index System (CODIS) and National DNA Index System (NDIS) Programs
URLs: http://www.fbi.gov/hq/lab/codis/index1.htm (CODIS), http://foia.fbi.gov/dna552.htm (NDIS)
Phone: (202) 324-3000
J. Edgar Hoover Building
935 Pennsylvania Avenue, NW
Washington, DC 20535-0001
National database of DNA profiles from convicted criminals and crime scenes (NDIS) and distributed database and software that coordinates local, state, and national forensic DNA databases (CODIS). Both were authorized by the DNA Identification Act of 1994 and went into operation in 1998.

Federation of American Societies for Experimental Biology (FASEB)
E-mail: webmaster@faseb.org
URL: http://www.faseb.org
Phone: (301) 634-7000
9650 Rockville Pike
Bethesda, MD 20814
Coalition of societies that promote advancement in biomedical sciences and provide educational meetings and publications to disseminate biological research results. Advocates for the interests of biomedical and life sciences in matters of public policy.

Food and Agriculture Organization of the United Nations (FAO)
URL: http://www.fao.org
E-mail: FAO-HQ@fao.org
Phone: (39) 065-7054243
Viale delle Terme di Caracalla
00100 Rome
Italy
The United Nations's lead agency for agriculture, forestry, fisheries, and rural development, the FAO aims to raise levels of nutrition and standards of living, improve agricultural productivity, and better the condition of rural populations. It has provided some advice on regulation of genetically modified crops and foods but does not believe that such foods are essential to alleviate world hunger. It has an electronic forum on biotechnology in food and agriculture at http://www.fao.org/biotech/index.asp?lang=en.

Food and Drug Administration (FDA)
URL: http://www.fda.gov
Phone: (888) 463-6332
5600 Fishers Lane
Rockville, MD 20857-0001
U.S. government agency that regulates food, including genetically modified foods, and drugs and medical treatments, including drugs made by biotechnology and gene therapy. Web site has pages on bioengineered food and gene therapy.

Food First/Institute for Food and Development Policy
URL: http://www.foodfirst.org
E-mail: foodfirst@foodfirst.org
Phone: (510) 654-4400
398 60th Street
Oakland, CA 94618
This think tank highlights root causes and value-based solutions to world hunger and poverty. Opposes control of agricultural biotechnology and food supply by large multinational corporations.

Foundation on Economic Trends (FOET)
URL: http://www.foet.org
E-mail: office@foet.org
Phone: (202) 466-2823
1660 L Street, NW
Suite 216
Washington, DC 20036
Headed by Jeremy Rifkin, this nonprofit foundation examines emerging trends in science and technology and their impacts on society, culture, the economy, and the environment. Areas of interest include transgenic animals, patents on living things, and human gene alteration. The group urges caution in use of genetic technology.

Friends of the Earth
URL: http://www.foe.org/
E-mail: foe@foe.org
Phone: (877) 843-8687
1717 Massachusetts Avenue, NW
Suite 600
Washington, DC 20036-2002
Part of an international federation of environmental organizations (Friends of the Earth International) that try to work toward sustainable societies, protect the environment (including biological diversity), and promote justice and equal access to resources and opportunities for all the world's people. Among other things, the group fears the effects of multinational corporations' control of genetically modified crops on farmers in developing nations.

Future Generations
URL: http://www.eugenics.net
E-mail: vancourt@eugenics.net
Strives to leave a legacy of good health, high intelligence, and noble character to future generations through voluntary "humanitarian eugenics." Web site includes papers explaining and defending the organization's point of view.

Genetic Alliance
URL: http://www.geneticalliance.org
E-mail: information@geneticalliance.org

Phone: (202) 966-5557
4301 Connecticut Avenue, NW
Suite 404
Washington, DC 20008-2369
International coalition of consumer, health professional, and patient advocacy organizations working to improve the quality of life of everyone affected by genetics, especially people with genetic conditions and their families.

Genetic Engineering Action Network (GEAN)
URL: http://www.geaction.org/index.html
E-mail: info@geaction.org
Phone: (617) 661-6626
11 Ward Street, Suite 200
Somerville, MA 02143
Network of grassroots activists, nongovernmental organizations, farmer groups, academics, and scientists concerned about the risks to the environment, biodiversity, and human health, and the socioeconomic and ethical consequences, of biotechnology and genetic engineering.

Genetic Interest Group (GIG)
URL: http://www.gig.org.uk
E-mail: post@gig.org.uk
Phone: (44) 0207 704 3141
Unit 4d, Leroy House
436 Essex Road
London N1 3QP
United Kingdom
Alliance of British support groups for people and families with inherited diseases or diseases with a significant genetic component. Among other things, the group works to prevent discrimination based on the misapplication of genetic information, to promote recognition of the health benefits that can result from genetic research, and to guarantee that such research ultimately benefits individual patients.

Genetically Engineered Food Alert (GE Food Alert)
URL: http://www.gefoodalert.org/pages/home.cfm
Phone: (800) 390-3373
1200 18th Street, NW
Fifth Floor
Washington, DC 20036
Coalition of seven organizations united in their commitment to testing and labeling genetically engineered food. It wages campaigns against companies that use genetically modified foods and provides educational material on the dangers of genetically altered crops.

Genetics and Public Policy Center
URL: http://www.dnapolicy.org
E-mail: inquiries@DNApolicy.org
Phone: (202) 663-5971
1717 Massachusetts Avenue, NW
Suite 530
Washington, DC 20036
The center is funded by the Pew Charitable Trusts and is part of the Berman Bioethics Institute at Johns Hopkins University. It aims to be an independent and objective source of credible information

on genetic technologies and policies for the public, media, and policy makers. It focuses on reproductive genetics. It conducts polls, produces reports, sponsors conferences, and performs detailed legal and policy analyses on the subject.

Genetics Society of America (GSA)
URL: http://www.genetics-gsa.org
E-mail: estrass@genetics-gsa.org
Phone: (301) 634-7300
9650 Rockville Pike
Bethesda, MD 20814-3998
Professional society of scientists and academicians working in the field of genetic studies. Publishes monthly journal, *Genetics*, and educational/career materials.

GeneWatch UK
URL: http://www.genewatch.org
E-mail: mail@genewatch.org
Phone: (44) 01298 871898
The Mill House
Manchester Road
Tideswell, Buxton
Derbyshire SK17 8LN
United Kingdom
Public interest group that aims to ensure that genetic technologies are developed and used in ways that promote human health, protect the environment, and respect the rights of humans and interests of animals. It addresses GM crops and food, GM animals, human genetics, laboratory use, biological weapons, and patenting, focusing on the risks and downsides of genetic technology.

Greenpeace
URL: http://www.greenpeaceusa.org
Phone: (800) 326-0959
702 H Street, NW
Washington, DC 20001
Environmental group whose aims include protection of global biodiversity, requirement of careful regulation and labeling of genetically engineered foods, and opposition to patenting of living things. Believes that genetically engineered foods pose environmental and health risks.

Hastings Center
URL: http://www.thehastingscenter.org
Phone: (845) 424-4040
Route 9-D
21 Malcolm Gordon Road
Garrison, NY 10524-5555
The center addresses fundamental ethical issues in health, medicine, and the environment, including issues related to biotechnology and human genetics. Publishes bimonthly journal, *The Hastings Center Report*, a study of ethical issues related to research on human subjects, and other papers.

America's Health Insurance Plans (AHIP)
URL: http://www.aahp.org/template.cfm
1601 Pennsylvania Avenue, NW
South Building, Suite 500
Washington, DC 20004

Advocacy group for the health insurance industry. Opposes legislation that restricts insurers' use of genetic information.

Human Cloning Foundation
URL: http://www.humancloning.org
Stresses the positive aspects of human cloning and promotes education, awareness, and research about human cloning and other biotechnology. Web site offers papers supporting human cloning and lists of books about cloning, and genetic engineering.

Human Fertilisation and Embryology Authority (HFEA)
URL: http://www.hfea.gov.uk/Home
E-mail: admin@hfea.gov.uk
Phone: (44) 020 7377 5077
Paxton House
30 Artillery Lane
London E1 7LS
United Kingdom
Governmental body that regulates, licenses, and collects data on fertility treatments and human embryo research in Britain. Provides detailed information and advice to the public on these subjects.

Human Genetics Alert (HGA)
URL: http://www.hgalert.org
E-mail: info@hgalert.org
Phone: (44) (0) 207 704 6100
Unit 112, Aberdeen House
22–24 Highbury Grove
London N5 2EA
United Kingdom
Independent public interest watchdog group funded by a leading British charity. Informs the public and recommends policies about human genetics issues. Does not object to genetic research in general, but does oppose genetic discrimination, human cloning, and germ-line gene modification. Is concerned about "new eugenics," genetic determinism, use of genetic technologies to exacerbate social inequalities, and influence of commercial motives and sociocultural attitudes on genetic research.

Human Genetics Commission (HGC)
URL: http://www.hgc.gov.uk
E-mail: hgc@doh.gsi.gov.uk
Phone: (+44) 020 7972 1518
Department of Health
Area 652C, Skipton House
80 London Road
London SE1 6LH
United Kingdom
Advises the British government on ways that new developments in human genetics may affect people and health care. Issues reports and other publications.

Human Genome Organization (HUGO)
URL: http://www.gene.ucl.ac.uk/hugo
E-mail: hugo@hugo-international.org
Phone: (44) 207 935 8085
144 Harley Street
London W1G 7LD
United Kingdom
A leading international (chiefly European) professional organization

for scientists who study the human genome. Publishes a quarterly newsletter, *Genome Digest.*

The Innocence Project
URL: http://www.innocenceproject.org
E-mail: info@innocenceproject.org
100 Fifth Avenue
3rd Floor
New York, NY 10011
Founded by Barry Scheck and Peter Neufeld at the Cardozo Law School in New York in 1992, the project uses DNA tests and other evidence to show that certain people convicted of crimes are innocent and obtain their release.

Institute for Agriculture and Trade Policy (IATP)
URL: http://www.iatp.org
E-mail: iatp@iatp.org
Phone: (612) 870-0453
2105 First Avenue South
Minneapolis, MN 55404
Nonprofit research and education organization aimed at creating environmentally and economically sustainable communities and regions through sound agriculture and trade policy. Web site contains many essays on farming and trade, including some on genetically modified crops.

The Institute for Genomic Research (TIGR)
URL: http://www.tigr.org
Phone: (301) 838-0200
9712 Medical Center Drive
Rockville, MD 20850
TIGR is a nonprofit research institute that analyzes the genomes and gene products of a wide variety of organisms, from viruses to humans.

International Center for Technology Assessment (CTA)
URL: http://www.icta.org/aboutus/index.htm
E-mail: info@icta.org
Phone: (202) 547-9359
666 Pennsylvania Avenue, SE
Suite 302
Washington, DC 20003
Analyzes impacts of technology on society. Concerns include limiting genetic engineering, halting the patenting of life, and defending the integrity of food. Sponsors the Campaign for Food Safety and Biotechnology Watch Human Applications.

International Centre for Genetic Engineering and Biotechnology (ICGEB)
URL: http://www.icgeb.trieste.it
E-mail: kerbav@icgeb.org
Phone: (39) 040 37571
AREA Science Park
Padriciano 99
34012 Trieste
Italy
Established by the United Nations Industrial Development Organization in 1995 to provide research, training, and scientific services in the safe use of molecular biology and biotechnology, especially as it applies to the needs of developing nations.

International Cloning Society
URL: http://www.angelfire.com/la/jfled/ics.html
E-mail: adlafferty@msn.com
Acts as agency to preserve cell/DNA specimens of people who want to be cloned in the future, particularly for purposes of space travel.

International Food Information Council Foundation (IFIC)
URL: http://ific.org/
E-mail: foodinfo@ific.org
Phone: (202) 296-6540
1100 Connecticut Avenue, NW
Suite 430
Washington, DC 20036
Communicates science-based information on food safety and nutrition to health professionals, media, and others providing information to consumers. Supported by the food, beverage, and agricultural industries. Has information on food biotechnology.

International Service for the Acquisition of Agri-Biotech Applications (ISAAA)
URL: http://www.isaaa.org
E-mail: americenter@isaaa.org
Phone: (607) 255-1724
ISAAA AmeriCenter
417 Bradfield Hall
Cornell University
Ithaca, NY 14853
Nonprofit organization that delivers the benefits of new agricultural biotechnologies to the poor in developing countries. It aims to share these powerful technologies with those who stand to benefit from them and at the same time establish an enabling environment for their safe use. The group has centers in the United States, Kenya, and the Philippines.

International Society for Environmental Biotechnology (ISEB)
URL: http://cape.uwaterloo.ca/research/iseb/iseb.htm
E-mail: iseb@cape.uwaterloo.ca
Phone: (519) 746-4979 (Fax)
Department of Chemical Engineering
University of Waterloo
Waterloo, Ontario N2L 3G1
Canada
Interdisciplinary communication network of scientists and others interested in environmental biotechnology, or the development, use, and regulation of biological systems for remediation of contaminated environments and for environment-friendly processes.

Kennedy Institute of Ethics, Georgetown University
URL: http://www.georgetown.edu/research/kie/site/index.htm
Phone: (202) 687-8099
Healy, 4th Floor
Georgetown University
Washington, DC 20057
This teaching and research center sponsors research on medical ethics and related policy issues, including issues related to human genetics and gene alteration. Publishes a

journal, newsletter, a bibliography, and other materials.

March of Dimes Birth Defects Federation
URL: http://www.modimes.org
1275 Mamaroneck Avenue
White Plains, NY 10605
Works to prevent and treat birth defects, including those caused by genetic mutations. Offers educational material on genetic and other birth defects and reports on such subjects as newborn screening programs.

Mothers for Natural Law
URL: http://www.safe-food.org
P.O. Box 1177
Fairfield, IA 52556
Opposes genetic engineering, especially genetically engineered food.

National Agricultural Biotechnology Council (NABC)
URL: http://www.cals.cornell.edu/extension/nabc
E-mail: NABC@cornell.edu
Phone: (607) 254-4856
419 Boyce Thompson Institute
Tower Road
Ithaca, NY 14853
Provides an open forum to discuss issues related to agricultural biotechnology and encourage the field's safe, ethical, efficacious, and equitable development. Composed of major nonprofit agricultural biotechnology research and/or teaching institutions in Canada and the United States. Offers reports of annual meetings, papers, and a newsletter, *NABC News*.

National Center for Biotechnology Information (NCBI)
URL: http://www.ncbi.nlm.nih.gov
E-mail: info@ncbi.nlm.nih.gov
Phone: (301) 496-2475
National Library of Medicine
Building 38A
Bethesda, MD 20894
Offers information about biotechnology, including a newsletter, *NCBI News*, and genetic databases and analysis software.

National Center for Food and Agricultural Policy (NCFAP)
URL: http://www.ncfap.org
E-mail: ncfap@ncfap.org
Phone: (202) 328-5048
1616 P Street, NW
First Floor
Washington, DC 20036
Private nonprofit research organization that conducts studies in biotechnology, pesticides, international trade and development, and farm and food policy. Although officially a nonadvocacy group, the organization favors biotechnology, and several news stories claim that it is associated with the biotechnology and grocery industries.

National Corn Growers Association (NCGA)
URL: http://www.ncga.com
E-mail: corninfo@ncga.com
Phone: (636) 733-9004

632 Cepi Drive
Chesterfield, MO 63005
Corn growers' trade association. Believes that genetically altered corn holds great promise for corn farmers but would like to see the use of such crops spread more slowly. Warns that acceptance of genetically modified corn as food depends on better methods of informing consumers and on better management and regulation of crops.

National Environmental Trust (NET)
URL: http://environet.policy.net
E-mail: netinfo@environet.org
Phone: (202) 887-8800
1200 18th Street, NW
Fifth Floor
Washington, DC 20036
Nonprofit, nonpartisan group that aims to educate citizens about environmental problems and how they affect human health and quality of life. It opposes genetically modified foods.

National Family Farm Coalition (NFFC)
URL: http://www.nffc.net
E-mail: nffc@nffc.net
Phone: (202) 543-5675
110 Maryland Avenue, NE
Suite 307
Washington, DC 20002
Links grassroots organizations working on family farm issues. Strongly opposes genetically modified crops. Conducts Farmer to Farmer Campaign on Genetic Engineering, calling for sustainable production and consumers' right to GMO-free food.

National Farmers Union (NFU)
URL: http://www.nfu.org
Phone: (800) 347-1961
11900 East Cornell Avenue
Aurora, CO 80014-3194
Protects and enhances the economic interests and quality of life of family farmers and ranchers. Believes that farmers must make individual decisions about whether to plant genetically modified crops.

National Food Processors Association (NFPA)
URL: http://www.nfpa-food.org
E-mail: info@nfpa-food.org
Phone: (202) 639-5900
1350 I Street, NW
Suite 300
Washington, D.C. 20005
Food processing industry trade association. The group supports U.S. protests to the World Trade Organization about European refusal to accept GM crops, but it opposes engineering crops to produce pharmaceuticals unless regulators can guarantee that such crops will not enter the food supply.

National Human Genome Research Institute (NHGRI)
URL: http://www.genome.gov/
Phone: (301) 402-0911
Communications and Public Liaison Branch
National Human Genome Research Institute
National Institutes of Health

Building 31A, Room 4B09
31 Center Drive, MSC 2152
9000 Rockville Pike
Bethesda, MD 20892-2152
Part of the National Institutes of Health, the NHGRI heads the Human Genome Project. It offers reports and databases of genetic sequencing and other research, including links to other groups doing genome research. Its Ethical, Legal, and Social Implications (ELSI) Working Group has studied such issues as privacy and fairness in use of genetic information. It offers reports and fact sheets on its work.

National Institutes of Health Office of Biotechnology Activities (NIH OBA)
URL: http://www4.od.nih.gov/oba
E-mail: oba@od.nih.gov
Phone: (301) 496-9838
6705 Rockledge Drive
Suite 750, MSC 7985
Bethesda, MD 20892-7985
U.S. governmental agency that monitors scientific progress in human genetics research in order to anticipate future developments, including ethical, legal, and social concerns; manages the operation of, and provides analytical support to, the NIH Recombinant DNA Advisory Committee (RAC) and two committees that advise the Secretary of Health and Human Services on genetic issues; coordinates and provides liaison with federal and nongovernmental national and international organizations concerned with recombinant DNA, human gene transfer, genetic technologies, and xenotransplantation; provides advice to the NIH director, other federal agencies, and state regulatory organizations concerning genetic technologies; and develops and implements NIH policies and procedures for the safe conduct of recombinant DNA activities and human gene transfer.

National Newborn Screening and Genetics Resource Center (NNSGRC)
URL: http://genes-r-us.uthscsa.edu
E-mail: therrell@uthscsa.edu (Brad Therrell)
Phone: (512) 454-6419
1912 West Anderson Lane
Suite 210
Austin, TX 78757
Formed by a cooperative agreement between the federal government's Maternal and Child Health Bureau and the University of Texas. Provides a forum for interaction among consumers, health care professionals, researchers, organizations, and policy makers in refining and developing public health newborn screening and genetics programs. Serves as a national resource center for information and education in the areas of newborn screening and genetics. Stresses states' role in ensuring effective and nondiscriminatory genetic testing of newborns.

National Organization for Rare Disorders (NORD)
URL: http://www.rarediseases.org
E-mail: orphan@rarediseases.org

Phone: (800) 999-6673
55 Kenosia Avenue
P.O. Box 1968
Danbury, CT 06813-1968
A federation of voluntary health organizations dedicated to helping people with rare ("orphan") diseases and to assisting the organizations that serve them. Offers several databases online.

National Society of Genetic Counselors (NSGC)
URL: http://www.nsgc.org
E-mail: FYI@nsgc.org
Phone: (610) 872-7608
233 Canterbury Drive
Wallingford, PA 19086-6617
Professional organization of genetic counselors. Web site includes consumer information and press releases.

Nuffield Council on Bioethics
URL: http://www.nuffieldbioethics.org/home
E-mail: bioethics@nuffieldfoundation.org
Phone: (44) 020 7681 9619
28 Bedford Square
London WC1B 3JS
United Kingdom
Established by the Trustees of the Nuffield Foundation in 1991 to identify, examine, and report on the ethical questions raised by recent advances in biological and medical research. Publishes papers on such subjects as genetically modified foods, stem cell research, and patents on living things.

Organic Consumers Association (OCA)
URL: http://www.organicconsumers.org
E-mail: staff@organicconsumers.org
Phone: (218) 226-4164
6101 Cliff Estate Road
Little Marais, MN 55614
Grassroots nonprofit organization that represents the interests of organic consumers in dealing with crucial issues of food safety, industrial agriculture, genetic engineering, corporate accountability, and environmental sustainability. Opposes genetically engineered food, human cloning, and patenting of living things.

Organic Trade Association (OTA)
URL: http://www.ota.com/index.html
E-mail: info@ota.com
Phone: (413) 774-7511
P.O. Box 547
Greenfield, MA 01302
Industry association of organic farmers. Calls for a moratorium on growing GMOs and mandatory labeling of existing ones. Opposes Bt crops because they may lead to insect resistance to one of the few pesticides that organic farmers can use.

Organisation for Economic Co-operation and Development (OECD)
URL: http://www.oecd.org/home
E-mail: icgb@oecd.org
Phone: (33) 1 4524-8200

2, rue André Pascal
F-75775 Paris Cedex 16
France
Includes 30 member countries that share a commitment to democratic government and the market economy. The group has relationships with some 70 other countries, nongovernmental organizations, and civil society. It is well known for its publications and statistics on a variety of economic and social issues, and governments often rely on its advice. Its Internal Co-ordination Group for Biotechnology has considerable information on this subject at http://www.oecd.org/topic/0,2686,en_2649_37437_1_1_1_1_37437,00.html.

Pesticide Action Network of North America (PANNA)
URL: http://www.panna.org
E-mail: panna@panna.org
Phone: (415) 981-1771
49 Powell Street
Suite 500
San Francisco, CA 94102
Works to replace pesticide use with ecologically sound and socially just alternatives. Opposes genetically modified foods.

Pew Initiative on Food and Biotechnology
URL: http://pewagbiotech.org
E-mail: inquiries@pewagbiotech.org
Phone: (202) 347-9044
1331 H Street, NW
Suite 900
Washington, DC 20005
Aims to be an independent and objective source of credible information about agricultural biotechnology for the public, media, and policy makers. Sponsors workshops and conferences and produces reports in order to encourage debate on the subject.

President's Council on Bioethics
URL: http://www.bioethics.gov
E-mail: info@bioethics.gov
Phone: (202) 296-4669
1801 Pennsylvania Avenue, NW
Suite 700
Washington, DC 20006
Seventeen-member council of renowned bioethicists and other scientists that advises the president of the United States on bioethical issues arising from advances in biomedical science and technology. In July 2002 it produced a highly publicized report that evaluated the ethics of human reproductive and research cloning and made recommendations for policy.

Research Foundation for Science, Technology and Ecology (RFSTE)
URL: http://www.vshiva.net
E-mail: rfste@vsnl.com, vshiva@vsnl.com
Phone: (91) 11-26968077
A-60, Hauz Khas
New Delhi 110016
India
Headed by activist Vandana Shiva, this group works to conserve biodiversity, protect the rights of indigenous peoples, and oppose centralized

systems of monoculture in agriculture, forestry, and fisheries. Shiva opposes patents on living things and "biopiracy," or exploitation of native plants, animals, and cultural knowledge by international biotechnology corporations.

Soil Association
URL: http://www.soilassociation.org/web/sa/saweb.nsf?Open
E-mail: info@soilassociation.org
Phone: (44) 0117 929 0661
Bristol House
40-56 Victoria Street
Bristol, BS1 6BY
United Kingdom
Supports organic food and farming and sustainable forestry. Opposes genetically modified crops and claims that farmers lose money on them.

Trans Atlantic Consumer Dialogue (TACD)
URL: http://www.tacd.org
E-mail: tacd@consint.org
Phone: (44) 207 226 66 63
TACD Secretariat
Consumers International, Office for Developed Economies
24 Highbury Crescent
London N5 1RX
United Kingdom
Forum of U.S. and European Union (EU) consumer organizations that develops joint consumer policy recommendations to the U.S. government and EU to promote consumers' interests in EU and U.S. policy making. In late 2002 it reproached the United States for failing to establish mandatory labeling and safety reviews for GM crops and praised the EU for its new traceability and labeling requirements and for promoting the "precautionary principle" in trade.

Union of Concerned Scientists (UCS)
URL: http://www.ucsusa.org/index.cfm
E-mail: ucs@ucsusa.org
Phone: (617) 547-5552
Two Brattle Square
Cambridge, MA 02238-9105
Works to improve the environment and protect human health, safety, and quality of life. Urges strong government regulation of agricultural biotechnology.

U.K. Forum for Genetics and Insurance (UKFGI)
URL: http://www.ukfgi.org.uk
E-mail: ukfgi@actuaries.org
Phone: (44) 020 7632 2177
Staple Inn Hall, High Holborn
London WC1V 7QJ
United Kingdom
Sponsored chiefly by the insurance industry but also includes other businesses, caring professions, charities, scientific organizations, national and local governments, and consumer groups. Aims to bring together medical and statistical research on the extra risks to people with conditions to which there is a significant genetic predisposition, to consider the value of the results of genetic tests in the assessment of people's insurability,

and to encourage further research and discussion in these areas.

U.S. Public Interest Research Group (U.S. PIRG)
URL: http://www.uspirg.org
E-mail: uspirg@pirg.org
Phone: (202) 546-9707
218 D Street, SE
Washington, DC 20003
The state PIRGs created the U.S. PIRG to act as a watchdog for the public interest in our nation's capital, much as the state PIRGs have worked to safeguard the public interest in state capitals. Aims to influence national policy by means of investigative research, media exposés, grassroots organizing, advocacy, and litigation. Opposes agricultural biotechnology.

World Intellectual Property Organization (WIPO)
URL: http://www.wipo.int
E-mail: publicinf@wipo.int
Phone: (41) 22 338 91 11
34, chemin des Colombettes
Geneva
Switzerland
WIPO is developing global services, units, and initiatives to respond efficiently and rapidly to the existing and emerging needs of its member states, in particular those flowing from the burst of knowledge-based economy. It seeks to provide an integrated approach to protecting intellectual property and promoting creativity, an alternative that will be more favored by developing countries than the WTO/GATT/TRIPS approach.

PART III

APPENDICES

APPENDIX A

Buck v. Bell, 274 U.S. 200 (1927)

1. The Virginia statute providing for the sexual sterilization of inmates of institutions supported by the State who shall be found to be afflicted with an hereditary form of insanity or imbecility, is within the power of the State under the Fourteenth Amendment. P. 207.

2. Failure to extend the provision to persons outside the institutions named does not render it obnoxious to the Equal Protection Clause. P. 208.

143 Va. 310, Affirmed.

Opinions

Error to a judgment of the Supreme Court of Appeals of the State of Virginia which affirmed a judgment ordering the Superintendent of the State Colony of Epileptics and Feeble Minded to perform the operation of salpingectomy on Carrie Buck, the plaintiff in error.

Holmes, J., Opinion of the Court

Mr. JUSTICE HOLMES delivered the opinion of the Court.

This is a writ of error to review a judgment of the Supreme Court of Appeals of the State of Virginia affirming a judgment of the Circuit Court of Amherst County by which the defendant in error, the superintendent of the State Colony for Epileptics and Feeble Minded, was ordered to perform the operation of salpingectomy upon Carrie Buck, the plaintiff in error, for the purpose of making her sterile. [143 Va. 310.] The case comes here upon the contention that the statute authorizing the judgment is void under the Fourteenth Amendment as denying to the plaintiff in error due process of law and the equal protection of the laws.

Carrie Buck is a feeble minded white woman who was committed to the State Colony above mentioned in due form. She is the daughter of a feeble minded mother in the same institution, and the mother of an illegitimate feeble minded child. She was eighteen years old at the time of the trial of her case in the Circuit Court, in the latter part of 1924. An Act of Virginia, approved March 20, 1924, recites that the health of the patient and the welfare of society may be promoted in certain cases by the sterilization of mental defectives, under careful safeguard, &c.; that the sterilization may be effected in males by vasectomy and in females by salpingectomy, without serious pain or substantial danger to life; that the Commonwealth is supporting in various institutions many defective persons who, if now discharged, would become a menace, but, if incapable of procreating, might be discharged with safety and become self-supporting with benefit to themselves and to society, and that experience has shown that heredity plays an important part in the transmission of insanity, imbecility, &c. The statute then enacts that, whenever the superintendent of certain institutions, including the above-named State Colony, shall be of opinion that it is for the best interests of the patients and of society that an inmate under his care should be sexually sterilized, he may have the operation performed upon any patient afflicted with hereditary forms of insanity, imbecility, &c., on complying with the very careful provisions by which the act protects the patients from possible abuse.

The superintendent first presents a petition to the special board of directors of his hospital or colony, stating the facts and the grounds for his opinion, verified by affidavit. Notice of the petition and of the time and place of the hearing in the institution is to be served upon the inmate, and also upon his guardian, and if there is no guardian, the superintendent is to apply to the Circuit Court of the County to appoint one. If the inmate is a minor, notice also is to be given to his parents, if any, with a copy of the petition. The board is to see to it that the inmate may attend the hearings if desired by him or his guardian. The evidence is all to be reduced to writing, and, after the board has made its order for or against the operation, the superintendent, or the inmate, or his guardian, may appeal to the Circuit Court of the County. The Circuit Court may consider the record of the board and the evidence before it and such other admissible evidence as may be offered, and may affirm, revise, or reverse the order of the board and enter such order as it deems just. Finally any party may apply to the Supreme Court of Appeals, which, if it grants the appeal, is to hear the case upon the record of the trial in the Circuit Court, and may enter such order as it thinks the Circuit Court should have entered. There can be no doubt that, so far as procedure is concerned, the rights of the patient are most carefully considered, and, as every step in this case was taken in scrupulous compliance with the statute and after months of observation, there is no doubt that, in that respect, the plaintiff in error has had due process of law.

The attack is not upon the procedure, but upon the substantive law. It seems to be contended that in no circumstances could such an order be justified. It certainly is contended that the order cannot be justified upon the existing grounds. The judgment finds the facts that have been recited, and that Carrie Buck is the probable potential parent of socially inadequate offspring, likewise afflicted, that she may be sexually sterilized without detriment to her general health, and that her welfare and that of society will be promoted by her sterilization, and thereupon makes the order. In view of the general declarations of the legislature and the specific findings of the Court, obviously we cannot say as matter of law that the grounds do not exist, and, if they exist, they justify the result. We have seen more than once that the public welfare may call upon the best citizens for their lives. It would be strange if it could not call upon those who already sap the strength of the State for these lesser sacrifices, often not felt to be such by those concerned, in order to prevent our being swamped with incompetence. It is better for all the world if, instead of waiting to execute degenerate offspring for crime or to let them starve for their imbecility, society can prevent those who are manifestly unfit from continuing their kind. The principle that sustains compulsory vaccination is broad enough to cover cutting the Fallopian tubes. *Jacobson v. Massachusetts*, 197 U.S. 11. Three generations of imbeciles are enough.

But, it is said, however it might be if this reasoning were applied generally, it fails when it is confined to the small number who are in the institutions named and is not applied to the multitudes outside. It is the usual last resort of constitutional arguments to point out shortcomings of this sort. But the answer is that the law does all that is needed when it does all that it can, indicates a policy, applies it to all within the lines, and seeks to bring within the lines all similarly situated so far and so fast as its means allow. Of course, so far as the operations enable those who otherwise must be kept confined to be returned to the world, and thus open the asylum to others, the equality aimed at will be more nearly reached.

APPENDIX B

DIAMOND V. CHAKRABARTY, 447 U.S. 303 (1980)

DIAMOND, COMMISSIONER OF PATENTS AND TRADEMARKS V. CHAKRABARTY CERTIORARI TO THE UNITED STATES COURT OF CUSTOMS AND PATENT APPEALS. NO. 79-136

Argued March 17, 1980
Decided June 16, 1980

Title 35 U.S.C. 101 provides for the issuance of a patent to a person who invents or discovers "any" new and useful "manufacture" or "composition of matter." Respondent filed a patent application relating to his invention of a human-made, genetically engineered bacterium capable of breaking down crude oil, a property which is possessed by no naturally occurring bacteria. A patent examiner's rejection of the patent application's claims for the new bacteria was affirmed by the Patent Office Board of Appeals on the ground that living things are not patentable subject matter under 101. The Court of Customs and Patent Appeals reversed, concluding that the fact that micro-organisms are alive is without legal significance for purposes of the patent law.

Held:

1. A live, human-made micro-organism is patentable subject matter under 101. Respondent's micro-organism constitutes a "manufacture" or "composition of matter" within that statute. Pp. 308–318.

(a) In choosing such expansive terms as "manufacture" and "composition of matter," modified by the comprehensive "any," Congress contemplated that the patent laws should be given wide scope, and the relevant legislative history also supports a broad construction. While laws of nature, physical phenomena, and abstract ideas are not patentable, respondent's claim is not to a hitherto unknown natural phenomenon, but to a nonnaturally occurring manufacture or composition of matter—a product of human ingenuity "having a distinctive name, character [and] use." *Hartranft v. Wiegmann*, 121 U.S. 609, 615. *Funk Brothers Seed Co. v. Kalo Inoculant Co.*, 333 U.S. 127, distinguished. Pp. 308–310.

(b) The passage of the 1930 Plant Patent Act, which afforded patent protection to certain asexually reproduced plants, and the 1970 Plant Variety Protection Act, which authorized protection for certain sexually reproduced plants but excluded bacteria from its protection, does not evidence congressional understanding that the terms "manufacture" or "composition of matter" in 101 do not include living things. Pp. 310–314. [447 U.S. 303, 304].

(c) Nor does the fact that genetic technology was unforeseen when Congress enacted 101 require the conclusion that micro-organisms cannot qualify as patentable subject matter until Congress expressly authorizes such protection. The unambiguous language of 101 fairly embraces respondent's invention. Arguments against patentability under 101, based on potential hazards that may be generated by genetic research, should be addressed to the Congress and the Executive, not to the Judiciary. Pp. 314–318.

596 F.2d 952, affirmed.

BURGER, C. J., delivered the opinion of the Court, in which STEWART, BLACKMUN, REHNQUIST, and STEVENS, JJ., joined. BRENNAN, J., filed a dissenting opinion, in which WHITE, MARSHALL, and POWELL, JJ., joined, post, p. 318.

Deputy Solicitor General Wallace argued the cause for petitioner. With him on the brief were Solicitor General McCree, Assistant Attorney General Shenefield, Harriet S. Shapiro, Robert B. Nicholson, Frederic Freilicher, and Joseph F. Nakamura.

Edward F. McKie, Jr., argued the cause for respondent. With him on the brief were Leo I. MaLossi, William E. Schuyler, Jr., and Dale H. Hoscheit.

MR. CHIEF JUSTICE BURGER delivered the opinion of the Court.

We granted certiorari to determine whether a live, human-made micro-organism is patentable subject matter under 35 U.S.C. 101.

I

In 1972, respondent Chakrabarty, a microbiologist, filed a patent application, assigned to the General Electric Co. The application asserted 36 claims related to Chakrabarty's invention of "a bacterium from the genus *Pseudomonas* containing therein at least two stable energy-generating plasmids, each of said plasmids providing a separate hydrocarbon degradative pathway." This human-made, genetically engineered bacterium is capable of breaking down multiple components of crude oil. Because of this property, which is possessed by no naturally occurring bacteria, Chakrabarty's invention is believed to have significant value for the treatment of oil spills.

Chakrabarty's patent claims were of three types: first, process claims for the method of producing the bacteria; [447 U.S. 303, 306] second, claims for an inoculum comprised of a carrier material floating on water, such as straw, and the new bacteria; and third, claims to the bacteria themselves. The patent examiner allowed the claims falling into the first two categories, but rejected claims for the bacteria. His decision rested on two grounds: (1) that micro-organisms are "products of nature," and (2) that as living things they are not patentable subject matter under 35 U.S.C. 101.

Chakrabarty appealed the rejection of these claims to the Patent Office Board of Appeals, and the Board affirmed the examiner on the second ground. Relying on the legislative history of the 1930 Plant Patent Act, in which Congress extended patent protection to certain asexually reproduced plants, the Board concluded that 101 was not intended to cover living things such as these laboratory created micro-organisms.

The Court of Customs and Patent Appeals, by a divided vote, reversed on the authority of its prior decision in *In re Bergy*, 563 F.2d 1031, 1038 (1977), which held that "the fact that microorganisms . . . are alive . . . [is] without legal significance" for purposes of the patent law. Subsequently, we granted the Acting Commissioner of Patents and Trademarks' petition for certiorari in *Bergy*, vacated the judgment, and remanded the case "for further consideration in light of *Parker v. Flook*, 437 U.S. 584 (1978)" 438 U.S. 902 (1978). The Court of Customs and Patent Appeals then vacated its judgment in *Chakrabarty* and consolidated the case with *Bergy* for reconsideration. After re-examining both cases in the light of our holding in *Flook*, that court, with one dissent, reaffirmed its earlier judgments 596 F.2d 952 (1979). [447 U.S. 303, 307].

The Commissioner of Patents and Trademarks again sought certiorari, and we granted the writ as to both *Bergy* and *Chakrabarty*. 444 U.S. 924 (1979). Since then, *Bergy* has been dismissed as moot, 444 U.S. 1028 (1980), leaving only *Chakrabarty* for decision.

II

The Constitution grants Congress broad power to legislate to "promote the Progress of Science and useful Arts, by securing for limited Times to Authors and Inventors the exclusive Right to their respective Writings and Discoveries." Art. I, 8, cl. 8. The patent laws promote this progress by offering inventors exclusive rights for a limited period as an incentive for their inventiveness and research efforts. *Kewanee Oil Co. v. Bicron Corp.*, 416 U.S. 470, 480–481 (1974); *Universal Oil Co. v. Globe Co.*, 322 U.S. 471, 484 (1944). The authority of Congress is exercised in the hope that "[t]he productive effort thereby fostered will have a positive effect on society through the introduction of new products and processes of manufacture into the economy, and the emanations by way of increased employment and better lives for our citizens." *Kewanee*, supra, at 480.

The question before us in this case is a narrow one of statutory interpretation requiring us to construe 35 U.S.C. 101, which provides:

"Whoever invents or discovers any new and useful process, machine, manufacture, or composition of matter, or any new and useful improvement thereof, may obtain a patent therefor, subject to the conditions and requirements of this title."

Specifically, we must determine whether respondent's micro-organism constitutes a "manufacture" or "composition of matter" within the meaning of the statute [447 U.S. 303, 308].

III

In cases of statutory construction we begin, of course, with the language of the statute. *Southeastern Community College v. Davis*, 442 U.S. 397, 405 (1979). And "unless otherwise defined, words will be interpreted as taking their ordinary, contemporary, common meaning." *Perrin v. United States*, 444 U.S. 37, 42 (1979). We have also cautioned that courts "should not read into the patent laws limitations and conditions which the legislature has not expressed." *United States v. Dubilier Condenser Corp.*, 289 U.S. 178, 199 (1933).

Guided by these canons of construction, this Court has read the term "manufacture" in 101 in accordance with its dictionary definition to mean "the production of articles for use from raw or prepared materials by giving to these materials new forms, qualities, properties, or combinations, whether by hand-labor or by machinery." *American Fruit Growers, Inc. v. Brogdex Co.*, 283 U.S. 1, 11 (1931). Similarly, "composition of matter" has been construed consistent with its common usage to include "all compositions of two or more substances and . . . all composite articles, whether they be the results of chemical union, or of mechanical mixture, or whether they be gases, fluids, powders or solids." *Shell Development Co. v. Watson*, 149 F. Supp. 279, 280 (DC 1957) (citing 1 A. Deller, *Walker on Patents* 14, p. 55 (1st ed. 1937)). In choosing such expansive terms as "manufacture" and "composition of matter," modified by the comprehensive "any," Congress plainly contemplated that the patent laws would be given wide scope.

The relevant legislative history also supports a broad construction. The Patent Act of 1793, authored by Thomas Jefferson, defined statutory subject matter as "any new and useful art, machine, manufacture, or composition of matter, or any new or useful improvement [thereof]." Act of Feb. 21, 1793, 1, 1 Stat. 319. The Act embodied Jefferson's philosophy that "ingenuity should receive a liberal encouragement." [447 U.S. 303, 309] (*Writings of Thomas Jefferson* 75–76) (Washington ed. 1871). See *Graham v. John Deere Co.*, 383 U.S. 1, 7–10 (1966). Subsequent patent statutes in 1836, 1870, and 1874 employed this same broad language. In 1952, when the patent laws were recodified, Congress replaced the word "art" with "process," but otherwise left Jefferson's language intact. The Committee Reports accompanying the 1952 Act inform us that Congress intended statutory subject matter to "include anything under the sun that is made by man." S. Rep. No. 1979, 82d Cong., 2d Sess., 5 (1952); H. R. Rep. No. 1923, 82d Cong., 2d Sess., 6 (1952).

This is not to suggest that 101 has no limits or that it embraces every discovery. The laws of nature, physical phenomena, and abstract ideas have been held not patentable. See *Parker v. Flook*, 437 U.S. 584 (1978); *Gottschalk v. Benson*, 409 U.S. 63, 67 (1972); *Funk Brothers Seed Co. v. Kalo Inoculant Co.*, 333 U.S. 127, 130 (1948); *O'Reilly v. Morse*, 15 How. 62, 112–121 (1854); *Le Roy v. Tatham*, 14 How. 156, 175 (1853). Thus, a new mineral discovered in the earth or a new plant found in the wild is not patentable subject matter. Likewise, Einstein could not patent his celebrated law that $E=mc^2$.; nor could Newton have patented the law of gravity. Such discoveries are "manifestations of . . . nature, free to all men and reserved exclusively to none." *Funk*, supra, at 130.

Judged in this light, respondent's micro-organism plainly qualifies as patentable subject matter. His claim is not to a hitherto unknown natural phenomenon, but to a nonnaturally occurring manufacture or composition of matter—a product of human ingenuity "having a distinctive name, character [and] [447 U.S. 303, 310] use" *Hartranft v. Wiegmann*, 121 U.S. 609, 615 (1887). The point is underscored dramatically by comparison of the invention here with that in *Funk*. There, the patentee had discovered that there existed in nature certain species of root-nodule bacteria which did not exert a mutually inhibitive effect on each other. He used that discovery to produce a mixed culture capable of inoculating the seeds of leguminous plants. Concluding that the patentee had discovered "only some of the handiwork of nature," the Court ruled the product nonpatentable:

"Each of the species of root-nodule bacteria contained in the package infects the same group of leguminous plants which it always infected. No species acquires a different use. The combination of species produces no new bacteria, no change in the six species of bacteria, and no enlargement of the range of their utility. Each species has the same effect it always had. The bacteria perform in their natural way. Their use in combination does not improve in any way their natural functioning. They serve the ends nature originally provided and act quite independently of any effort of the patentee" [333 U.S., at 131].

Here, by contrast, the patentee has produced a new bacterium with markedly different characteristics from any found in nature and one having the potential for significant utility. His discovery is not nature's handiwork, but his own; accordingly it is patentable subject matter under 101.

IV

Two contrary arguments are advanced, neither of which we find persuasive.

(A)

The petitioner's first argument rests on the enactment of the 1930 Plant Patent Act, which afforded patent protection to certain asexually reproduced plants, and the 1970 Plant [447 U.S. 303, 311] Variety Protection Act, which authorized protection for certain sexually reproduced plants but excluded bacteria from its protection. In the petitioner's view, the passage of these Acts evidences congressional understanding that the terms "manufacture" or "composition of matter" do not include living things; if they did, the petitioner argues, neither Act would have been necessary.

We reject this argument. Prior to 1930, two factors were thought to remove plants from patent protection. The first was the belief that plants, even those artificially bred, were products of nature for purposes of the patent law. This position appears to have derived from the decision of the Patent Office in *Ex parte* Latimer, 1889 Dec. Com. Pat. 123, in which a patent claim for fiber found in the needle of the *Pinus australis* was rejected. The Commissioner reasoned that a contrary result would permit "patents [to] be obtained upon the trees of the forest and the plants of the earth, which of course would be unreasonable and impossible" Id., at 126. The Latimer case, it seems, came to "se[t] forth the general stand taken in these matters" that plants were natural products not subject to patent protection. Thorne, *Relation of Patent Law to Natural Products*, 6 J. Pat. Off. Soc. 23, 24 [447 U.S. 303, 312] (1923). The second obstacle to patent protection for plants was the fact that plants were thought not amenable to the "written description" requirement of the patent law. See 35 U.S.C. 112. Because new plants may differ from old only in color or perfume, differentiation by written description was often impossible. See Hearings on H. R. 11372 before the House Committee on Patents, 71st Cong., 2d Sess., 7 (1930) (memorandum of Patent Commissioner Robertson).

In enacting the Plant Patent Act, Congress addressed both of these concerns. It explained at length its belief that the work of the plant breeder "in aid of nature" was patentable invention. S. Rep. No. 315, 71st Cong., 2d Sess., 6–8 (1930); H. R. Rep. No. 1129, 71st Cong., 2d Sess., 7–9 (1930). And it relaxed the written description requirement in favor of "a description . . . as complete as is reasonably possible" 35 U.S.C. 162. No Committee or Member of Congress, however, expressed the broader view, now urged by the petitioner, that the terms "manufacture" or "composition of matter" exclude living things. The sole support for that position in the legislative history of the 1930 Act is found in the conclusory statement of Secretary of Agriculture Hyde, in a letter to the Chairmen of the House and Senate Committees considering the 1930 Act, that "the patent laws . . . at the present time are understood to cover only inventions or discoveries in the field of inanimate nature." See S. Rep. No. 315, supra, at Appendix A; H. R. Rep. No. 1129, supra, at Appendix A. Secretary Hyde's opinion, however, is not entitled to controlling weight. His views were solicited on the administration of the new law and not on the scope of patentable [447 U.S. 303, 313] subject matter—an area beyond his competence. Moreover, there is language in the House and Senate Committee Reports suggesting that to the extent Congress considered the matter it found the Secretary's dichotomy unpersuasive. The Reports observe:

"There is a clear and logical distinction between the discovery of a new variety of plant and of certain inanimate things, such, for example, as a new and useful natural mineral. The mineral is created wholly by nature unassisted by man. . . . On the other hand, a plant discovery resulting from cultivation is unique, isolated, and is not repeated by nature, nor can it be reproduced by nature unaided by man. . . ." S. Rep. No. 315, supra, at 6; H. R. Rep. No. 1129, supra, at 7.

Congress thus recognized that the relevant distinction was not between living and inanimate things, but between products of nature, whether living or not, and human-made inventions. Here, respondent's micro-organism is the result of human ingenuity and research. Hence, the passage of the Plant Patent Act affords the Government no support.

Nor does the passage of the 1970 Plant Variety Protection Act support the Government's position. As the Government acknowledges, sexually reproduced plants were not included under the 1930 Act because new varieties could not be reproduced true-to-type through seedlings. Brief for Petitioner 27, n. 31. By 1970, however, it was generally recognized that true-to-type reproduction was possible and that plant patent protection was therefore appropriate. The 1970 Act extended that protection. There is nothing in its language or history to suggest that it was enacted because 101 did not include living things.

In particular, we find nothing in the exclusion of bacteria from plant variety protection to support the petitioner's position. See n. 7, supra. The legislative history gives no reason for this exclusion. As the Court of Customs and [447 U.S. 303, 314] Patent Appeals suggested, it may simply reflect congressional agreement with the result reached by that court in deciding *In re Arzberger*, 27 C. C. P. A. (Pat.) 1315, 112 F.2d 834 (1940), which held that bacteria were not plants for the purposes of the 1930 Act. Or it may reflect the fact that prior to 1970 the Patent Office had issued patents for bacteria under 101. In any event, absent some clear indication that Congress "focused on [the] issues . . . directly related to the one presently before the Court," *SEC v. Sloan*, 436 U.S. 103, 120–121 (1978), there is no basis for reading into its actions an intent to modify the plain meaning of the words found in 101. See *TVA v. Hill*, 437 U.S. 153, 189–193 (1978); *United States v. Price*, 361 U.S. 304, 313 (1960).

(B)

The petitioner's second argument is that micro-organisms cannot qualify as patentable subject matter until Congress expressly authorizes such protection. His position rests on the fact that genetic technology was unforeseen when Congress enacted 101. From this it is argued that resolution of the patentability of inventions such as respondent's should be left to Congress. The legislative process, the petitioner argues, is best equipped to weigh the competing economic, social, and scientific considerations involved, and to determine whether living organisms produced by genetic engineering should receive patent protection. In support of this position, the petitioner relies on our recent holding in *Parker v. Flook*, 437 U.S. 584 (1978), and the statement that the judiciary "must proceed cautiously when . . . asked to extend [447 U.S. 303, 315] patent rights into areas wholly unforeseen by Congress." [Id., at 596].

It is, of course, correct that Congress, not the courts, must define the limits of patentability; but it is equally true that once Congress has spoken it is "the province and duty of the judicial department to say what the law is." *Marbury v. Madison*, 1 Cranch 137, 177 (1803). Congress has performed its constitutional role in defining patentable subject matter in 101; we perform ours in construing the language Congress has employed. In so doing, our obligation is to take statutes as we find them, guided, if ambiguity appears, by the legislative history and statutory purpose. Here, we perceive no ambiguity. The subject-matter provisions of the patent law have been cast in broad terms to fulfill the constitutional and statutory goal of promoting "the Progress of Science and the useful Arts" with all that means for the social and economic benefits envisioned by Jefferson. Broad general language is not necessarily ambiguous when congressional objectives require broad terms.

Nothing in *Flook* is to the contrary. That case applied our prior precedents to determine that a "claim for an improved method of calculation, even when tied to a specific end use, is unpatentable subject matter under 101." [437 U.S., at 595, n. 18]. The Court carefully scrutinized the claim at issue to determine whether it was precluded from patent protection under "the principles underlying the prohibition against patents for 'ideas' or phenomena of nature." [Id., at 593]. We have done that here. *Flook* did not announce a new principle that inventions in areas not contemplated by Congress when the patent laws were enacted are unpatentable *per se*.

To read that concept into *Flook* would frustrate the purposes of the patent law. This Court frequently has observed that a statute is not to be confined to the "particular application[s] . . . contemplated by the legislators." *Barr v. United States*, 324 U.S. 83, 90 (1945). Accord, *Browder v. United States*, 312 U.S. 335, 339 (1941); *Puerto Rico v. Shell Co.*, [447 U.S. 303, 316] 302 U.S. 253, 257 (1937). This is especially true in the field of patent law. A rule that unanticipated inventions are without protection would conflict with the core concept of the patent law that anticipation undermines patentability. See *Graham v. John Deere Co.*, 383 U.S., at 12–17. Mr. Justice Douglas reminded that the inventions most benefiting mankind are those that "push back the frontiers of chemistry, physics, and the like." *Great A. & P. Tea Co. v. Supermarket Corp.*, 340 U.S. 147, 154 (1950) (concurring opinion). Congress employed broad general language in drafting 101 precisely because such inventions are often unforeseeable.

To buttress his argument, the petitioner, with the support of amicus, points to grave risks that may be generated by research endeavors such as respondent's. The briefs present a gruesome parade of horribles. Scientists, among them Nobel laureates, are quoted suggesting that genetic research may pose a serious threat to the human race, or, at the very least, that the dangers are far too substantial to permit such research to proceed apace at this time. We are told that genetic research and related technological developments may spread pollution and disease, that it may result in a loss of genetic diversity, and that its practice may tend to depreciate the value of human life. These arguments are forcefully, even passionately, presented; they remind us that, at times, human ingenuity seems unable to control fully the forces it creates—that, with Hamlet, it is sometimes better "to bear those ills we have than fly to others that we know not of."

It is argued that this Court should weigh these potential hazards in considering whether respondent's invention is [447 U.S. 303, 317] patentable subject matter under 101. We disagree. The grant or denial of patents on micro-organisms is not likely to put an end to genetic research or to its attendant risks. The large amount of research that has already occurred when no researcher had sure knowledge that patent protection would be available suggests that legislative or judicial fiat as to patentability will not deter the scientific mind from probing into the unknown any more than Canute could command the tides. Whether respondent's claims are patentable may determine whether research efforts are accelerated by the hope of reward or slowed by want of incentives, but that is all.

What is more important is that we are without competence to entertain these arguments—either to brush them aside as fantasies generated by fear of the unknown, or to act on them. The choice we are urged to make is a matter of high policy for resolution within the legislative process after the kind of investigation, examination, and study that legislative bodies can provide and courts cannot. That process involves the balancing of competing values and interests, which in our democratic system is the business of elected representatives. Whatever their validity, the contentions now pressed on us should be addressed to the political branches of the Government, the Congress and the Executive, and not to the courts. [447 U.S. 303, 318].

We have emphasized in the recent past that "[o]ur individual appraisal of the wisdom or unwisdom of a particular [legislative] course . . . is to be put aside in the process of interpreting a statute." *TVA v. Hill*, 437 U.S., at 194. Our task, rather, is the narrow one of determining what Congress meant by the words it used in the statute; once that is done our powers are exhausted. Congress is free to amend 101 so as to exclude from patent protection organisms produced by genetic engineering. Cf. 42 U.S.C. 2181 (a), exempting from patent protection inventions "useful solely in the utilization of special nuclear material or atomic energy in an atomic weapon." Or it may choose to craft a statute specifically designed for such living things. But, until Congress takes such action, this Court must construe the language of 101 as it is. The language of that section fairly embraces respondent's invention.

Accordingly, the judgment of the Court of Customs and Patent Appeals is Affirmed. . . .

MR. JUSTICE BRENNAN, with whom MR. JUSTICE WHITE, MR. JUSTICE MARSHALL, and MR. JUSTICE POWELL join, dissenting.

I agree with the Court that the question before us is a narrow one. Neither the future of scientific research, nor even the ability of respondent Chakrabarty to reap some monopoly profits from his pioneering work, is at stake. Patents on the processes by which he has produced and employed the new living organism are not contested. The only question we need decide is whether Congress, exercising its authority under Art. I, 8, of the Constitution, intended that he be able to secure a monopoly on the living organism itself, no matter how produced or how used. Because I believe the Court has misread the applicable legislation, I dissent. [447 U.S. 303, 319]

The patent laws attempt to reconcile this Nation's deep-seated antipathy to monopolies with the need to encourage progress. *Deepsouth Packing Co. v. Laitram Corp.*, 406 U.S. 518, 530–531 (1972); *Graham v. John Deere Co.*, 383 U.S. 1, 7–10 (1966). Given the complexity and legislative nature of this delicate task, we must be careful to extend patent protection no further than Congress has provided. In particular, were there an absence of legislative direction, the courts should leave to Congress the decisions whether and how far to extend the patent privilege into areas where the common understanding has been that patents are not available. Cf. *Deepsouth Packing Co. v. Laitram Corp.*, supra.

In this case, however, we do not confront a complete legislative vacuum. The sweeping language of the Patent Act of 1793, as re-enacted in 1952, is not the last pronouncement Congress has made in this area. In 1930 Congress enacted the Plant Patent Act affording patent protection to developers of certain asexually reproduced plants. In 1970 Congress enacted the Plant Variety Protection Act to extend protection to certain new plant varieties capable of sexual reproduction. Thus, we are not dealing—as the Court would have it—with the routine problem of "unanticipated inventions." Ante, at 316. In these two Acts Congress has addressed the general problem of patenting animate inventions and has chosen carefully limited language granting protection to some kinds of discoveries, but specifically excluding others. These Acts strongly evidence a congressional limitation that excludes bacteria from patentability. [447 U.S. 303, 320]

First, the Acts evidence Congress' understanding, at least since 1930, that 101 does not include living organisms. If newly developed living organisms not naturally occurring had been patentable under 101, the plants included in the scope of the 1930 and 1970 Acts could have been patented without new legislation. Those plants, like the bacteria involved in this case, were new varieties not naturally occurring. Although the Court, ante, at 311, rejects this line of argument, it does not explain why the Acts were necessary unless to correct a pre-existing situation. I cannot share the Court's implicit assumption that Congress was engaged in either idle exercises or mere correction of the public record when it enacted the 1930 and 1970 Acts. And Congress certainly thought it was doing something significant. The Committee Reports contain expansive prose about the previously unavailable benefits to be derived from extending patent protection to plants. H. R. [447 U.S. 303, 321] Rep. No. 91-1605, pp. 1–3 (1970); S. Rep. No. 315, 71st Cong., 2d Sess., 1–3 (1930). Because Congress thought it had to legislate in order to make agricultural "human-made inventions" patentable and because the legislation Congress enacted is limited, it follows that Congress never meant to make items outside the scope of the legislation patentable.

Second, the 1970 Act clearly indicates that Congress has included bacteria within the focus of its legislative concern, but not within the scope of patent protection. Congress specifically excluded bacteria from the coverage of the 1970 Act. 7 U.S.C. 2402 (a). The Court's attempts to supply explanations for this explicit exclusion ring hollow. It is true that there is not mention in the legislative history of the exclusion, but that does not give us license to invent reasons. The fact is that Congress, assuming that animate objects as to which it had not specifically legislated could not be patented, excluded bacteria from the set of patentable organisms.

The Court protests that its holding today is dictated by the broad language of 101, which cannot "be confined to the 'particular application[s] . . . contemplated by the legislators.'" Ante, at 315, quoting *Barr v. United States*, 324 U.S. 83, 90 (1945). But as I have shown, the Court's decision does not follow the unavoidable implications of the statute. Rather, it extends the patent system to cover living material [447 U.S. 303, 322] even though Congress plainly has legislated in the belief that 101 does not encompass living organisms. It is the role of Congress, not this Court, to broaden or narrow the reach of the patent laws. This is especially true where, as here, the composition sought to be patented uniquely implicates matters of public concern.

[footnotes omitted]

APPENDIX C

NORMAN-BLOODSAW V. LAWRENCE BERKELEY LABORATORY, 135 F.3D 1260 (1998)

Note: This case was in the United States Court of Appeals for the Ninth Circuit.

MARYA S. NORMAN-BLOODSAW; EULALIO R. FUENTES; VERTIS B. ELLIS; MARK E. COVINGTON; JOHN D. RANDOLPH; ADRIENNE L. GARCIA; and BRENDOLYN B. SMITH, Plaintiffs-Appellants, v. No. 96-16526 LAWRENCE BERKELEY LABORATORY; CHARLES V. SHANK, Director of D.C. No. Lawrence Berkeley Laboratory; CV-95-03220-VRW HENRY H. STAUFFER, M.D.; LISA SNOW, M.D.; T. F. BUDINGER, M.D.; WILLIAM G. DONALD, JR., M.D.; FEDERICO PENA, Secretary of the Department of Energy;* and THE REGENTS OF THE UNIVERSITY OF CALIFORNIA, a non-profit public corporation, Defendants-Appellees. Appeal from the United States District Court for the Northern District of California Vaughn R. Walker, District Judge, Presiding Argued and Submitted June 10, 1997—San Francisco, California Filed February 3, 1998

Before: Stephen Reinhardt, Thomas G. Nelson, and Michael Daly Hawkins, Circuit Judges. Opinion by Judge Reinhardt.

---

*Federico Pena has been substituted for his predecessor in office, Hazel O'Leary, pursuant to Fed. R. App. P. 43(c)(1). 1149

Appendix C

This appeal involves the question whether a clerical or administrative worker who undergoes a general employee health examination may, without his knowledge, be tested for highly private and sensitive medical and genetic information such as syphilis, sickle-cell trait, and pregnancy.

Lawrence Berkeley Laboratory is a research institution jointly operated by state and federal agencies. Plaintiffs-appellants, present and former employees of Lawrence, allege that in the course of their mandatory employment entrance examinations and on subsequent occasions, Lawrence, without their knowledge or consent, tested their blood and urine for intimate medical conditions—namely, syphilis, sickle-cell trait, and pregnancy. Their complaint asserts that this testing violated Title VII of the Civil Rights Act of 1964, the Americans with Disabilities Act (ADA), and their right to privacy as guaranteed by both the United States and State of California Constitutions. The district court granted the defendants-appellees' motions for dismissal, judgment on the pleadings, and summary judgment on all of plaintiffs-appellants' claims. We affirm as to the ADA claims, but reverse as to the Title VII and state and federal privacy claims.

BACKGROUND

Plaintiffs Marya S. Norman-Bloodsaw, Eulalio R. Fuentes, Vertis B. Ellis, Mark E. Covington, John D. Randolph, Adrienne L. Garcia, and Brendolyn B. Smith are current and former administrative and clerical employees of defendant Lawrence Berkeley Laboratory ("Lawrence"), a research facility operated by the appellee Regents of the University of California pursuant to a contract with the United States Department of Energy (the Department). Defendant Charles V. Shank is the director of Lawrence, and defendants Henry H. Stauffer, Lisa Snow, T. F. Budinger, and William G. Donald, Jr., are all current or former physicians in its medical department. The named defendants are sued in both their official and individual capacities.

The Department requires federal contractors such as Lawrence to establish an occupational medical program. Since 1981, it has required its contractors to perform "preplacement examinations" of employees as part of this program, and until 1995, it also required its contractors to offer their employees the option of subsequent "periodic health examinations." The mandatory preplacement examination occurs after the offer of employment but prior to the assumption of job duties. The Department actively oversees Lawrence's occupational health program, and, prior to 1992, specifically required syphilis testing as part of the preplacement examination.

With the exception of Ellis, who was hired in 1968 and underwent an examination after beginning employment, each of the plaintiffs received written offers of employment expressly conditioned upon a "medical examination," "medical approval," or "health evaluation." All accepted these offers and underwent preplacement examinations, and Randolph and Smith underwent subsequent examinations as well.

In the course of these examinations, plaintiffs completed medical history questionnaires and provided blood and urine samples. The questionnaires asked, inter alia, whether the patient had ever had any of sixty-one medical conditions, including "[s]ickle cell anemia," "[v]enereal disease," and, in the case of women, "[m]enstrual disorders."

The blood and urine samples given by all employees during their preplacement examinations were tested for syphilis; in addition, certain samples were tested for sickle-cell trait; and certain samples were tested for pregnancy. Lawrence discontinued syphilis testing in April 1993, pregnancy testing in December 1994, and sickle-cell trait testing in June 1995. Defendants assert that they discontinued syphilis testing because of its limited usefulness in screening healthy populations, and that they discontinued sickle-cell trait testing because, by that time, most African-American adults had already been tested at birth. Lawrence continues to perform pregnancy testing, but only on an optional basis. Defendants further contend that "for many years" signs posted in the health examination rooms and "more recently" in the reception area stated that the tests at issue would be administered.

Following receipt of a right-to-sue letter from the EEOC, plaintiffs filed suit in September 1995 on behalf of all past and present Lawrence employees who have ever been subjected to the medical tests at issue. Plaintiffs allege that the testing of their blood and urine samples for syphilis, sickle-cell trait, and pregnancy occurred without their knowledge or consent, and without any subsequent notification that the tests had been conducted. They also allege that only black employees were tested for sickle-cell trait and assert the obvious fact that only female employees were tested for pregnancy.

Finally, they allege that Lawrence failed to provide safeguards to prevent the dissemination of the test results. They contend that they did not discover that the disputed tests had been conducted until approximately January 1995, and specifically deny that they observed any signs indicating that such tests would be performed. Plaintiffs do not allege that the defendants took any subsequent employment-related action on the basis of their test results, or that their test results have been disclosed to third parties.

On the basis of these factual allegations, plaintiffs contend that the defendants violated the ADA by requiring, encouraging, or assisting in medical testing that was neither job-related nor consistent with business necessity. Second, they contend that the defendants violated the federal constitutional right to privacy by conducting the testing at issue, collecting and maintaining the results of the testing, and failing to provide adequate safeguards against disclosure of the results. Third, they contend that the testing violated their right to privacy under Article I, [section] 1 of the California Constitution. Finally, plaintiffs contend that Lawrence and the Regents violated Title VII by singling out black employees for sickle-cell trait testing and by performing pregnancy testing on female employees generally.

The state defendants moved for judgment on the pleadings or, in the alternative, for summary judgment. The sole federal defendant (the "Secretary"), then-Secretary of Energy Hazel O'Leary, moved to dismiss the various claims against her for lack of subject matter jurisdiction and for failure to state a claim. Turning first to the ADA claims, the district court reasoned that because the medical questionnaires inquired into information such as venereal disease and reproductive status, plaintiffs were on notice at the time of their examinations that Lawrence was engaging in medical inquiries that were neither job-related nor consistent with business necessity. Thus, given that the most recent examination occurred over two years before the filing of the complaint, the district court held that all of the ADA claims were time-barred. It also rejected the argument that storage of the test results constitutes a "continuing violation" of the ADA that tolls the limitations period.

The district court next concluded that the federal privacy claims were also time-barred and, in the alternative, failed on the merits. On the grounds that the tests were "part of a comprehensive medical examination to which plaintiffs had consented," and that plaintiffs had completed a medical history form of "highly personal questions" that included inquiries concerning "venereal disease," "sickle-cell anemia," and "menstrual problems," it concluded that plaintiffs were aware at the time of their examinations "of sufficient facts to put them on notice" that their blood and urine would be tested for syphilis, sickle-cell trait, and pregnancy, and that their claims were thus time-barred. The district court then held, in the alternative, that the testing had not violated plaintiffs' due process right to privacy. Relying again on the fact that the tests were performed as part of a general medical examination "that covered the same areas as the tests themselves," it concluded that any "additional incremental intrusion" from the tests was so minimal that no constitutional violation could have occurred despite defendants' failure to identify "an undisputed legitimate governmental purpose" for the tests.

Finally, the district court held that the Title VII claims, even if viable, were time-barred for the same reasons as were the privacy and ADA claims. It also concluded that plaintiffs had failed to state a cognizable Title VII claim, reasoning that plaintiffs had "neither alleged nor shown any connection between these discontinued confidential tests and [their] employment terms or conditions, either in the past or in the future"; and finding that "[p]laintiffs' charge of stigmatic harm, stripped of hyperbole, speculation, and conjecture . . . evaporates."

This appeal followed.

DISCUSSION

I. Statute of Limitations

[1] The district court dismissed all of the claims on statute of limitations grounds because it found that the limitations period began to run at the time the tests were taken, in which case each cause of action would be time-barred. Federal law determines when the limitations period begins to run, and the general federal rule is that "a limitations period begins to run when the plaintiff knows or has reason to know of the injury which is the basis of the action." *Trotter v. International Longshoremen's & Warehousemen's Union*, 704 F.2d 1141, 1143 (9th Cir. 1983). Because the district court resolved the statute of limitations question on summary judgment, we must determine, viewing all facts in the light most favorable to plaintiffs and resolving all factual ambiguities in their favor, whether the district court erred in determining that plaintiffs knew or should have known of the particular testing at issue when they underwent the examinations.

[2] We find that whether plaintiffs knew or had reason to know of the specific testing turns on material issues of fact that can only be resolved at trial. Plaintiffs' declarations clearly state that at the time of the examination they did not know that the testing in question would be performed, and they neither saw signs nor received any other indications to that effect. The district court had three possible reasons for concluding that plaintiffs knew or should have expected the tests at issue: (1) they submitted to an occupational preplacement examination; (2) they answered written questions as to whether they had had "venereal disease," "menstrual problems," or "sickle-cell anemia"; and (3) they voluntarily gave blood and urine samples. Given the present state of the record, these facts are hardly sufficient to establish that plaintiffs either knew or should have known that the particular testing would take place.

The question of what tests plaintiffs should have expected or foreseen depends in large part upon what preplacement medical examinations usually entail, and what, if anything, plaintiffs were told to expect. The record strongly suggests that plaintiffs' submission to the exam did not serve to afford them notice of the particular testing involved. The letters that plaintiffs received informed them merely that a "medical examination," "medical approval," or "health evaluation" was an express condition of employment.

These letters did not inform plaintiffs that they would be subjected to comprehensive diagnostic medical examinations that would inquire into intimate health matters bearing no relation to their responsibilities as administrative or clerical employees.

The record, indeed, contains considerable evidence that the manner in which the tests were performed was inconsistent with sound medical practice. Plaintiffs introduced before the district court numerous expert declarations by medical scholars roundly condemning Lawrence's alleged practices and explaining, inter alia, that testing for syphilis, sickle-cell trait, and pregnancy is not an appropriate part of an occupational medical examination and is rarely if ever done by employers as a matter of routine; that Lawrence lacked any reasonable medical or public health basis for performing these tests on clerical and administrative employees such as plaintiffs; and that the performance of such tests without explicit notice and informed consent violates prevailing medical standards.

The district court also appears to have reasoned that plaintiffs knew or had reason to know of the tests because they were asked questions on a medical form concerning "venereal disease," "sickle-cell anemia," and "menstrual disorders," and because they gave blood and urine samples. The fact that plaintiffs acquiesced in the minor intrusion of checking or not checking three boxes on a questionnaire does not mean that they had reason to expect further intrusions in the form of having their blood and urine tested for specific conditions that corresponded tangentially if at all to the written questions. First, the entries on the questionnaire were neither identical to nor, in some cases, even suggestive of the characteristics for which plaintiffs were tested. For example, sickle-cell trait is a genetic condition distinct from actually having sickle-cell anemia, and pregnancy is not considered a "menstrual disorder" or a "venereal disease." Second, and more important, it is not reasonable to infer that a person who answers a questionnaire upon personal knowledge is put on notice that his employer will take intrusive means to verify the accuracy of his answers. There is a significant difference between answering on the basis of what you know about your health and consenting to let someone else investigate the most intimate aspects of your life. Indeed, a reasonable person could conclude that by completing a written questionnaire, he has reduced or eliminated the need for seemingly redundant and even more intrusive laboratory testing in search of highly sensitive and non-job-related information.

Furthermore, if plaintiffs' evidence concerning reasonable medical practice is to be credited, they had no reason to think that tests would be performed without their consent simply because they had answered some questions on a form and had then, in addition, provided bodily fluid samples: Plaintiffs could reasonably have expected Lawrence to seek their consent before running any tests not usually performed in an occupational health exam—particularly tests for intimate medical conditions bearing no relationship to their responsibilities or working conditions as clerical employees. The mere fact that an employee has given a blood or urine sample does not provide notice that an employer will perform any and all tests on that specimen that it desires,—no matter how invasive—particularly where, as here, the employer has yet to offer a valid reason for the testing.

[3] In sum, the district court erred in holding as a matter of law that the plaintiffs knew or had reason to know of the nature of the tests as a result of their submission to the preemployment medical examinations. Because the question of what testing, if any, plaintiffs had reason to expect turns on material factual issues that can only be resolved at trial, summary judgment on statute of limitations grounds was inappropriate with respect to the causes of action based on an invasion of privacy in violation of the Federal and California Constitutions, and also on the Title VII claims.

II. Federal Constitutional Due Process Right of Privacy

The district court also ruled, in the alternative, on the merits of all of plaintiffs' claims except the ADA claims. We first examine its ruling with respect to the claim for violation of the federal constitutional right to privacy. While acknowledging that the government had failed to identify any "undisputed legitimate governmental purpose" for the three tests, the district court concluded that no violation of plaintiffs' right to privacy could have occurred because any intrusions arising from the testing were *de minimis* in light of (1) the "large overlap" between the subjects covered by the medical questionnaire and the three tests and (2) the "overall intrusiveness" of "a full-scale physical examination." We hold that the district court erred.

Because the ADA claims fail on the merits, as discussed below, we do not determine whether the district court erred in dismissing those claims on statute of limitations grounds.

[4] The constitutionally protected privacy interest in avoiding disclosure of personal matters clearly encompasses medical information and its confidentiality. *Doe v. Attorney General of the United States*, 941 F.2d 780, 795 (9th Cir. 1991) (citing *United States v. Westinghouse Elec. Corp.*, 638 F.2d 570, 577 (3d Cir. 1980)); *Roe v. Sherry*, 91 F.3d 1270, 1274 (9th Cir. 1996); see also *Doe v. City of New York*, 15 F.3d 264, 267–69 (2d Cir. 1994). Although cases defining the privacy interest in medical information have typically involved its disclosure to "third" parties, rather than the collection of information by illicit means, it goes without saying that the most basic violation possible involves the performance of unauthorized tests—that is, the non-consensual

retrieval of previously unrevealed medical information that may be unknown even to plaintiffs. These tests may also be viewed as searches in violation of Fourth Amendment rights that require Fourth Amendment scrutiny. The tests at issue in this case thus implicate rights protected under both the Fourth Amendment and the Due Process Clause of the Fifth or Fourteenth Amendments. *Yin v. California*, 95 F.3d 864, 870 (9th Cir. 1996), cert. denied, 117 S. Ct. 955 (1997).

[5] Because it would not make sense to examine the collection of medical information under two different approaches, we generally "analyze [medical tests and examinations] under the rubric of [the Fourth] Amendment. " Id. at 871 & n.12. Accordingly, we must balance the government's interest in conducting these particular tests against the plaintiffs' expectations of privacy. Id. at 873. Furthermore, "application of the balancing test requires not only considering the degree of intrusiveness and the state's interests in requiring that intrusion, but also 'the efficacy of this [the state's] means for meeting' its needs." Id. (quoting *Vernonia Sch. Dist. 47J v. Acton*, 515 U.S. 646, 660 (1995)).

[6] The district court erred in dismissing the claims on the ground that any violation was *de minimis*, incremental, or overlapping. The latter two grounds are actually just the court's explanations for its adoption of its "*de minimis*" conclusion. They are not in themselves reasons for dismissal. Nor if the violation is otherwise significant does it become insignificant simply because it is overlapping or incremental. We cannot, therefore, escape a scrupulous examination of the nature of the violation, although we can, of course, consider whether the plaintiffs have in fact consented to any part of the alleged intrusion.

[7] One can think of few subject areas more personal and more likely to implicate privacy interests than that of one's health or genetic make-up. Doe, 15 F.3d at 267 ("Extension of the right to confidentiality to personal medical information recognizes there are few matters that are quite so personal as the status of one's health"); see *Vernonia Sch. Dist. 47J*, 515 U.S. at 658 (noting under Fourth Amendment analysis that "it is significant that the tests at issue here look only for drugs, and not for whether the student is, for example, epileptic, pregnant, or diabetic"). Furthermore, the facts revealed by the tests are highly sensitive, even relative to other medical information. With respect to the testing of plaintiffs for syphilis and pregnancy, it is well established in this circuit "that the Constitution prohibits unregulated, unrestrained employer inquiries into personal sexual matters that have no bearing on job performance." *Schowengerdt v. General Dynamics Corp.*, 823 F.2d 1328, 1336 (9th Cir. 1987) (citing *Thorne v. City of El Segundo*, 726 F.2d 459, 470 (9th Cir. 1983)). The fact that one has syphilis is an intimate matter that pertains to

one's sexual history and may invite tremendous amounts of social stigma. Pregnancy is likewise, for many, an intensely private matter, which also may pertain to one's sexual history and often carries far-reaching societal implications. See *Thorne*, 726 F.2d at 468–70; Doe, 15 F.3d at 267 (noting discrimination and intolerance to which HIV-positive persons are exposed). Finally, the carrying of sickle-cell trait can pertain to sensitive information about family history and reproductive decisionmaking. Thus, the conditions tested for were aspects of one's health in which one enjoys the highest expectations of privacy.

[8] As discussed above, with respect to the question of the statute of limitations, there was little, if any, "overlap" between what plaintiffs consented to and the testing at issue here. Nor was the additional invasion only incremental. In some instances, the tests related to entirely different conditions. In all, the information obtained as the result of the testing was qualitatively different from the information that plaintiffs provided in their answers to the questions, and was highly invasive. That one has consented to a general medical examination does not abolish one's privacy right not to be tested for intimate, personal matters involving one's health—nor does consenting to giving blood or urine samples, or filling out a questionnaire. As we have made clear, revealing one's personal knowledge as to whether one has a particular medical condition has nothing to do with one's expectations about actually being tested for that condition. Thus, the intrusion was by no means de minimis. Rather, if unauthorized, the testing constituted a significant invasion of a right that is of great importance, and labelling it minimal cannot and does not make it so.

[9] Lawrence further contends that the tests in question, even if their intrusiveness is not de minimis, would be justified by an employer's interest in performing a general physical examination. This argument fails because issues of fact exist with respect to whether the testing at issue is normally part of a general physical examination. There would of course be no violation if the testing were authorized, or if the plaintiffs reasonably should have known that the blood and urine samples they provided would be used for the disputed testing and failed to object. However, as we concluded in Section I, material issues of fact exist as to those questions.

Summary judgment in the alternative on the merits of the federal constitutional privacy claim was therefore incorrect.

III. Right to Privacy Under Article I, [Section] 1 of the California Constitution

With respect to the state privacy claims, defendants argue, as they did with respect to the federal privacy claims, that the intrusions occasioned by the testing were so minimal that the government need not demonstrate a legitimate interest in performing the tests. In the alternative, they argue that the intrusions were so minimal that plaintiffs' privacy interests were necessarily overcome by the government's interest in performing the preplacement examinations. We understand this argument to be essentially the same as the argument that these tests are a part of an ordinary general medical examination. Defendants urge no additional governmental interest but appear to rely entirely on the interest that any employer might assert in requiring potential employees to undergo general medical testing. The district court did not adopt either of the defendants' positions expressly but simply ruled that plaintiffs "could not proceed" because the "undisputed facts"—namely, completion of the medical questionnaire, consent to the preplacement examination, and the voluntary giving of blood and urine samples—showed that the tests had inflicted "only a *de minimis* privacy invasion."

[10] To assert a cause of action under Article I, § 1 of the California Constitution, one must establish three elements: (1) a legally protected privacy interest; (2) a reasonable expectation of privacy under the circumstances; and (3) conduct by the defendant that amounts to a "serious invasion" of the protected privacy interest. *Loder v. City of Glendale*, 927 P.2d 1200, 1228 (Cal. 1997) (quoting *Hill v. National Collegiate Athletic Ass'n*, 865 P.2d 633, 657 (Cal. 1994), cert. denied, 118 S.Ct. 44 (1997)). These elements must be "viewed simply as 'threshold elements,'" after which the court must conduct a balancing test between the "countervailing interests" for the conduct in question and the intrusion on privacy resulting from the conduct. A showing of "countervailing interests" may, in turn, be rebutted by a showing that there were "feasible and effective alternatives" with a "lesser impact on privacy interests." *Hill*, 865 P.2d at 657.

[11] For much the same reasons as we have discussed above with respect to the statute of limitations and federal privacy claims, the district court erred in dismissing the state constitutional privacy claim. The only possible difference between the state claim and the federal claim is the threshold requirement that the invasion be serious, and for purposes of summary judgment, that requirement has been more than met.

For the reasons discussed above, we find that material issues of fact exist with respect to whether the defendants had any interest at all in obtaining the information and whether plaintiffs had a reasonable expectation of privacy under the circumstances. Both these questions involve a factual dispute regarding the ordinary or accepted medical practice regarding general or pre-employment medical exams.

Accordingly, the district court also erred in dismissing the state constitutional privacy claims.

IV. Title VII Claims

The district court also dismissed the Title VII counts on the merits on the grounds that plaintiffs had failed to state a claim because the "alleged classifications, standing alone, do not suffice to provide a cognizable basis for relief under Title VII" and because plaintiffs had neither alleged nor demonstrated how these classifications had adversely affected them.

[12] Section 703(a) of Title VII of the Civil Rights Act of 1964 provides that it is unlawful for any employer:

(1) to fail or refuse to hire or to discharge any individual, or otherwise to discriminate against any individual with respect to his compensation, terms, conditions, or privileges of employment, because of such individual's race, color, religion, sex, or national origin; or

(2) to limit, segregate, or classify his employees or applicants for employment in any way which would deprive or tend to deprive any individual of employment opportunities or otherwise adversely affect his status as an employee, because of such individual's race, color, religion, sex, or national origin.

42 U.S.C. § 2000e-2(a)

The Pregnancy Discrimination Act further provides that discrimination on the basis of "sex" includes discrimination "on the basis of pregnancy, childbirth, or related medical conditions." 42 U.S.C. § 2000e(k). "In accordance with Congressional intent, the above language is to be read in the broadest possible terms. The intent of Congress was not to list specific discriminatory practices, nor to definitively set out the scope of the activities covered." EEOC Compliance Manual (CCH) § 613.1, at P 2901 (citing *Rogers v. EEOC*, 454 F.2d 234 (5th Cir. 1971)).

Despite defendants' assertions to the contrary, plaintiffs' Title VII claims fall neatly into a Title VII framework: Plaintiffs allege that black and female employees were singled out for additional nonconsensual testing and that defendants thus selectively invaded the privacy of certain employees on the basis of race, sex, and pregnancy. The district court held that

(1) the tests did not constitute discrimination in the "terms" or "conditions" of plaintiffs' employment; and that (2) plaintiffs have failed to show any "adverse effect" as a result of the tests. It also granted the plaintiffs leave to amend their complaint to show adverse effect.

[13] Under [section] 2000e-2(a)(1), supra, an employer who "otherwise . . . discriminate[s]" with respect to the "terms" or "conditions" of employment on account of an illicit classification is subject to Title VII liability. It is well established that Title VII bars discrimination not only in the "terms" and "conditions" of ongoing employment, but also in the "terms" and "conditions" under which individuals may obtain employment. See, e.g., *Griggs*

v. Duke Power Co., 401 U.S. 424, 432–37 (1971) (facially neutral educational and testing requirements that are not reasonable measures of job performance and have disparate impact on hiring of minorities violate Title VII). Thus, for example, a requirement of preemployment health examinations imposed only on female employees, or a requirement of preemployment background security checks imposed only on black employees, would surely violate Title VII.

[14] In this case, the term or condition for black employees was undergoing a test for sickle-cell trait; for women it was undergoing a test for pregnancy. It is not disputed that the preplacement exams were, literally, a condition of employment: the offers of employment stated this explicitly. Thus, the employment of women and blacks at Lawrence was conditioned in part on allegedly unconstitutional invasions of privacy to which white and/or male employees were not subjected. An additional "term or condition" requiring an unconstitutional invasion of privacy is, without doubt, actionable under Title VII. Furthermore, even if the intrusions did not rise to the level of unconstitutionality, they would still be a "term" or "condition" based on an illicit category as described by the statute and thus a proper basis for a Title VII action. Thus, the district court erred in ruling on the leadings that the plaintiffs had failed to assert a proper Title VII claim under [section] 2000e-2(a)(1).

[15] The district court also erred in finding as a matter of law that there was no "adverse effect" with respect to the tests as required under [section] 2000e-2(a)(2). The unauthorized obtaining of sensitive medical information on the basis of race or sex would in itself constitute an "adverse effect," or injury, under Title VII.

Thus, it was error to rule that as a matter of law no "adverse effect" could arise from a classification that singled out particular groups for unconstitutionally invasive, non-consensual medical testing, and the district court erred in dismissing the Title VII claims on this ground as well.

V. The ADA Claims

Plaintiffs may challenge only the medical examinations that occurred "on or after January 26, 1992," which is the effective date of the ADA for public entities. The only plaintiffs who underwent any examinations or testing on or after that date are Fuentes and Garcia, who were tested in April 1992 and August 1993, respectively. The complaint alleges that defendants violated the ADA by requiring medical examinations and making medical inquiries that were "neither job-related nor consistent with business necessity." (Compl.P 64 (citing 42 U.S.C. [section] 12112(c)(4)). On appeal, plaintiffs also argue that "the ADA limits medical record keeping by an employer to the results of job-related examinations consistent with business necessity." Appellant Br. at 49 (citing 42 U.S.C.[section] 12112(d)). Plaintiffs do not allege that defendants made use of information gathered in the examinations to discriminate against them on the basis of disability; indeed, neither Garcia nor Fuentes received any positive test results.

[16] The ADA creates three categories of medical inquiries and examinations by employers: (1) those conducted prior to an offer of employment ("preemployment" inquiries and examinations); (2) those conducted "after an offer of employment has been made" but "prior to the commencement of . . . employment duties" ("employment entrance examinations"); and (3) those conducted at any point thereafter. It is undisputed that the second category, employment entrance examinations, as governed by [section] 12112(d)(3), are the examinations and inquiries to which Fuentes and Garcia were subjected. Unlike examinations conducted at any other time, an employment entrance examination need not be concerned solely with the individual's "ability to perform job-related functions," [section] 12112(d)(2); nor must it be "job-related or consistent with business necessity," [section] 12112(d)(4). Thus, the ADA imposes no restriction on the scope of entrance examinations; it only guarantees the confidentiality of the information gathered, [section] 12112(d)(3)(B), and restricts the use to which an employer may put the information. [section] 12112(d)(3)(C); see 42 U.S.C.[section] 12112(d)(1) (medical examinations and inquiries must be consistent with the general prohibition in [section] 12112(a) against discrimination on the basis of disability); 29 C.F.R. [section] 1630.14(b)(3) (if the results of the examination exclude an individual on the basis of disability, the exclusionary criteria themselves must be job-related and consistent with business necessity). Because the ADA does not limit the scope of such examinations to matters that are "job-related and consistent with business necessity," dismissal of the ADA claims was proper.

Plaintiffs' new argument on appeal that the ADA limits medical record-keeping to "the results of job-related examinations consistent with business necessity" also lacks merit. Section 12112(d)(3)(B) sets forth the conditions under which information obtained during the entrance examination must be kept but clearly does not purport to restrict the records that may be kept to matters that are "job-related and consistent with business necessity." Thus plaintiffs' ADA claims also fail in this respect.

The only possible ADA claim is directed at the defendants' alleged failure to maintain plaintiffs' medical records in the manner required by [section] 12112(d)(3)(B). The allegations in plaintiffs' complaint do not explicitly set forth such a violation but incorporate by reference the factual allegation that the defendants "[f]ail[ed] to provide safeguards to prevent the dissemination to third parties of sensitive medical information regarding the plaintiffs." On appeal the plaintiffs argue only that the defendants have "failed to describe the procedures by which a third party might gain access to the records, and the enforcement of any rules, policies, regulations or procedures to prevent third parties from gaining access to the records." To the extent that one can construe the complaint to allege that the defendants are in violation of [section] 12112(d)(3)(B), the bare allegation that defendants have not provided, or adequately described, safeguards fails to state a violation of the ADA requirements as set forth in [section] 12112(d)(3)(B) or as implemented in Department orders. See DOE Order 440.1 (Sep. 30, 1995); DOE Order 5480.8A (June 6, 1992); DOE Order 5480.8 (May 22, 1981). Accordingly, dismissal of the ADA claims was proper.

VI. Plaintiffs' Claims Are Not Moot

The Secretary contends that the claims against him in his official capacity for injunctive and declaratory relief are moot because (1) the only testing that the Department ever required was syphilis testing, and (2) the DOE order that required syphilis testing was cancelled on June 22, 1992, and replaced by a different order that requires "[u]rinalysis and serology" only "when indicated." Compare DOE Order 5480.8 (May 22, 1981), with DOE Order 5480.8A (June 26, 1992).

Although the state defendants do not raise the issue, a similar argument can be made on their behalf: Lawrence discontinued syphilis testing in April 1993, pregnancy testing in December 1994, and sickle-cell trait testing in June 1995.

[17] "[A] case is moot when the issues presented are no longer 'live' or the parties lack a legally cognizable interest in the outcome." *County of Los Angeles v. Davis*, 440 U.S.625, 631 (1979) [quoting *Powell v. McCormack*, 395 U.S. 486, 496 (1969)]. "Mere voluntary cessation of allegedly illegal conduct does not moot a case; if it did, the courts would be compelled to leave [t]he defendant . . . free to return to his old ways." *United States v. Concentrated Phosphate Export Ass'n*, 393 U.S. 199, 203 (1968) (quoting *United States v. W. T. Grant Co.*, 345 U.S. 629, 632 (1953)). Nevertheless, part or all of a case may become moot if (1) "subsequent events [have] made it absolutely clear that the allegedly wrongful behavior [cannot] reasonably be expected to recur," *Concentrated Phosphate*, 393 U.S. at 203, and (2) "interim relief or events have completely and irrevocably eradicated the effects of the alleged violation." *Lindquist v. Idaho State Bd. of Corrections*, 776 F.2d 851, 854 (9th Cir. 1985) (quoting *Davis*, 440 U.S. at 631). "The burden of demonstrating mootness 'is a heavy one.'" *Davis*, 440 U.S. at 631 (quoting *W. T. Grant*, 345 U.S. at 632–33).

[18] Defendants have not carried their heavy burden of establishing either that their alleged behavior cannot be reasonably expected to recur, or

that interim events have eradicated the effects of the alleged violation. First, they do not contend that the Department will never again require or permit, or that Lawrence will never again conduct, the tests at issue. They assert only that syphilis testing was discontinued because of its limited usefulness in screening healthy populations, and that sickle-cell trait testing was discontinued as redundant of testing that most African Americans now receive at birth. Moreover, in the case of pregnancy testing, they do not even argue that such testing is no longer medically useful; rather, they have simply made it optional. Defendants have neither asserted nor demonstrated that they will never resume mandatory testing for intimate medical conditions; nor have they offered any reason why they might not return in the future to their original views on the utility of mandatory testing. In contrast, plaintiffs have introduced evidence, in the form of correspondence between Lawrence and the department, that the syphilis tests were discontinued merely for reasons of "cost-effectiveness." See *Concentrated Phosphate*, 393 U.S. at 203 (holding that mere statement that it would be "uneconomical" for defendants to continue their allegedly wrongful conduct "cannot suffice to satisfy the heavy burden" of establishing mootness).

[19] Second, defendants also have not asserted that any "interim relief or events have completely and irrevocably eradicated the effects of the alleged violation." *Lindquist*, 776 F.2d at 854. Indeed, it is undisputed that the Department requires Lawrence to retain plaintiffs' test results and that Lawrence does in fact do so. See DOE Order 440.1, dated September 30, 1995 ("Employee medical records shall be adequately protected and stored permanently.") Even if the continued storage, against plaintiffs' wishes, of intimate medical information that was allegedly taken from them by unconstitutional means does not itself constitute a violation of law, it is clearly an ongoing "effect" of the allegedly unconstitutional and discriminatory testing, and expungement of the test results would be an appropriate remedy for the alleged violation. Cf. *Fendler v. United States Parole Comm'n*, 774 F.2d 975, 979 (9th Cir. 1985) ("Federal courts have the equitable power 'to order the expungement of Government records where necessary to vindicate rights secured by the Constitution or by statute.") [quoting *Chastain v. Kelley*, 510 F.2d 1232, 1235 (D.C. Cir. 1975)]; *Maurer v. Pitchess*, 691 F.2d 434, 437 (9th Cir. 1982). Accordingly, plaintiffs' claims for injunctive and declaratory relief are not moot.

VII. IRREPARABLE INJURY

[20] Finally, the Secretary contends that plaintiffs cannot seek injunctive relief because they have not alleged irreparable injury. To obtain injunctive relief, " '[a] reasonable showing' of a 'sufficient likelihood' that plaintiff will

be injured again is necessary." *Kruse v. State of Hawaii*, 68 F.3d 331, 335 (9th Cir. 1995) (internal quotation marks omitted); see *City of Los Angeles v. Lyons*, 461 U.S. 95, 111 (1983). "The likelihood of the injury recurring must be calculable and if there is no basis for predicting that any future repetition would affect the present plaintiffs, there is no case or controversy." *Sample v. Johnson*, 771 F.2d 1335, 1340 (9th Cir. 1985). In this case, plaintiffs seek not only to enjoin future illegal testing, but also to require defendants, inter alia, to notify all employees who may have been tested illegally; to destroy the results of such illegal testing upon employee request; to describe any use to which the information was put, and any disclosures of the information that were made; and to submit Lawrence's medical department to "independent oversight and monitoring."

[21] At the very least, the retention of undisputedly intimate medical information obtained in an unconstitutional and discriminatory manner would constitute a continuing "irreparable injury" for purposes of equitable relief. Moreover, the Department orders still require Lawrence to conduct preplacement examinations. DOE Order 440.1 (Sep. 30, 1995). Thus, there seems to be at least a reasonable possibility that Lawrence would again conduct undisclosed medical testing of its employees for intimate medical conditions. For these reasons, a request for injunctive relief is proper.

CONCLUSION

Because material and disputed issues of fact exist with respect to whether reasonable persons in plaintiffs' position would have had reason to know that the tests were being performed, and because the tests were a separate and more invasive intrusion into their privacy than the aspects of the examination to which they did consent, the district court erred in granting summary judgment on statute of limitations grounds with respect to the Title VII claims and the federal and state constitutional privacy claims. The district court also erred in dismissing the federal and state constitutional privacy claims and the Title VII claims on the merits. The district court's dismissal of the ADA claims was proper. None of the Secretary's arguments with respect to the claims brought against him in his official capacity has merit.

AFFIRMED IN PART, REVERSED IN PART, AND REMANDED.

[footnotes omitted]

APPENDIX D

BRAGDON V. ABBOTT 97 U.S. 156 (1998)

[portions are omitted]

ON WRIT OF CERTIORARI TO THE UNITED STATES COURT OF APPEALS FOR THE FIRST CIRCUIT

Justice Kennedy delivered the opinion of the Court.

We address in this case the application of the Americans with Disabilities Act of 1990 (ADA), 104 Stat. 327, 42 U.S.C. § 12101 et seq., to persons infected with the human immunodeficiency virus (HIV). We granted certiorari to review, first, whether HIV infection is a disability under the ADA when the infection has not yet progressed to the so-called symptomatic phase; and, second, whether the Court of Appeals, in affirming a grant of summary judgment, cited sufficient material in the record to determine, as a matter of law, that respondent's infection with HIV posed no direct threat to the health and safety of her treating dentist.

I

Respondent Sidney Abbott has been infected with HIV since 1986. When the incidents we recite occurred, her infection had not manifested its most serious symptoms. On September 16, 1994, she went to the office of petitioner Randon Bragdon in Bangor, Maine, for a dental appointment. She disclosed her HIV infection on the patient registration form. Petitioner completed a dental examination, discovered a cavity, and informed respondent of his policy against filling cavities of HIV-infected patients. He offered to perform the work at a hospital with no added fee for his services, though respondent would be responsible for the cost of using the hospital's facilities. Respondent declined.

Respondent sued petitioner under state law and §302 of the ADA, 104 Stat. 355, 42 U.S.C. § 12182 alleging discrimination on the basis of her disability.

The state law claims are not before us. Section 302 of the ADA provides:

"No individual shall be discriminated against on the basis of disability in the full and equal enjoyment of the goods, services, facilities, privileges, advantages, or accommodations of any place of public accommodation by any person who . . . operates a place of public accommodation." §12182(a).

The term "public accommodation" is defined to include the "professional office of a health care provider." §12181(7)(F).

A later subsection qualifies the mandate not to discriminate. It provides:

"Nothing in this subchapter shall require an entity to permit an individual to participate in or benefit from the goods, services, facilities, privileges, advantages and accommodations of such entity where such individual poses a direct threat to the health or safety of others." §12182(b)(3).

The United States and the Maine Human Rights Commission intervened as plaintiffs. After discovery, the parties filed cross-motions for summary judgment. The District Court ruled in favor of the plaintiffs, holding that respondent's HIV infection satisfied the ADA's definition of disability. 912 F. Supp. 580, 585–587 (Me. 1995). The court held further that petitioner raised no genuine issue of material fact as to whether respondent's HIV infection would have posed a direct threat to the health or safety of others during the course of a dental treatment. Id., at 587–591. The court relied on affidavits submitted by Dr. Donald Wayne Marianos, Director of the Division of Oral Health of the Centers for Disease Control and Prevention (CDC). The Marianos affidavits asserted it is safe for dentists to treat patients infected with HIV in dental offices if the dentist follows the so-called universal precautions described in the *Recommended Infection-Control Practices for Dentistry* issued by CDC in 1993 (1993 CDC *Dentistry Guidelines*). 912 F. Supp., at 589.

The Court of Appeals affirmed. It held respondent's HIV infection was a disability under the ADA, even though her infection had not yet progressed to the symptomatic stage. 107 F.3d 934, 939–943 (CA1 1997). The Court of Appeals also agreed that treating the respondent in petitioner's office would not have posed a direct threat to the health and safety of others. Id., at 943–948. Unlike the District Court, however, the Court of Appeals declined to rely on the Marianos affidavits. Id., at 946, n. 7. Instead the court relied on the 1993 CDC *Dentistry Guidelines*, as well as the *Policy on AIDS, HIV Infection and the Practice of Dentistry*, promulgated by the American Dental Association in 1991 (1991 American Dental Association *Policy on HIV*). 107 F.3d, at 945–946.

II

We first review the ruling that respondent's HIV infection constituted a disability under the ADA. The statute defines disability as:

"(A) a physical or mental impairment that substantially limits one or more of the major life activities of such individual;

"(B) a record of such an impairment; or

"(C) being regarded as having such impairment." §12102(2).

We hold respondent's HIV infection was a disability under subsection (A) of the definitional section of the statute. In light of this conclusion, we need not consider the applicability of subsections (B) or (C).

Our consideration of subsection (A) of the definition proceeds in three steps. First, we consider whether respondent's HIV infection was a physical impairment. Second, we identify the life activity upon which respondent relies (reproduction and child bearing) and determine whether it constitutes a major life activity under the ADA. Third, tying the two statutory phrases together, we ask whether the impairment substantially limited the major life activity. In construing the statute, we are informed by interpretations of parallel definitions in previous statutes and the views of various administrative agencies which have faced this interpretive question.

A

The ADA's definition of disability is drawn almost verbatim from the definition of "handicapped individual" included in the Rehabilitation Act of 1973, 29 U.S.C. § 706(8)(B) (1988 ed.), and the definition of "handicap" contained in the Fair Housing Amendments Act of 1988, 42 U.S.C. § 3602(h)(1) (1988 ed.). Congress' repetition of a well-established term carries the implication that Congress intended the term to be construed in accordance with pre-existing regulatory interpretations. See *FDIC v. Philadelphia Gear Corp.*, 476 U.S. 426, 437–438 (1986); *Commissioner v. Estate of Noel*, 380 U.S. 678, 681–682 (1965); *ICC v. Parker*, 326 U.S. 60, 65 (1945). In this case, Congress did more than suggest this construction; it adopted a specific statutory provision in the ADA directing as follows:

"Except as otherwise provided in this chapter, nothing in this chapter shall be construed to apply a lesser standard than the standards applied under title V of the Rehabilitation Act of 1973 (29 U.S.C. 790 *et seq.*) or the regulations issued by Federal agencies pursuant to such title." 42 U.S.C. § 12201(a).

The directive requires us to construe the ADA to grant at least as much protection as provided by the regulations implementing the Rehabilitation Act.

1. The first step in the inquiry under subsection (A) requires us to determine whether respondent's condition constituted a physical impairment. The Department of Health, Education and Welfare (HEW) issued the first regulations interpreting the Rehabilitation Act in 1977. The regulations are

of particular significance because, at the time, HEW was the agency responsible for coordinating the implementation and enforcement of §504. *Consolidated Rail Corporation v. Darrone*, 465 U.S. 624, 634, (1984) [citing Exec. Order No. 11914, 3 CFR 117 (1976–1980 Comp.)]. The HEW regulations, which appear without change in the current regulations issued by the Department of Health and Human Services, define "physical or mental impairment" to mean:

"(A) any physiological disorder or condition, cosmetic disfigurement, or anatomical loss affecting one or more of the following body systems: neurological; musculoskeletal; special sense organs; respiratory, including speech organs; cardiovascular; reproductive, digestive, genito-urinary; hemic and lymphatic; skin; and endocrine; or

"(B) any mental or psychological disorder, such as mental retardation, organic brain syndrome, emotional or mental illness, and specific learning disabilities." 45 CFR § 84.3(j)(2)(i) (1997).

In issuing these regulations, HEW decided against including a list of disorders constituting physical or mental impairments, out of concern that any specific enumeration might not be comprehensive. . . . [material establishing that asymptomatic HIV infection is a disability is omitted]

In light of the immediacy with which the virus begins to damage the infected person's white blood cells and the severity of the disease, we hold it is an impairment from the moment of infection. As noted earlier, infection with HIV causes immediate abnormalities in a person's blood, and the infected person's white cell count continues to drop throughout the course of the disease, even when the attack is concentrated in the lymph nodes. In light of these facts, HIV infection must be regarded as a physiological disorder with a constant and detrimental effect on the infected person's hemic and lymphatic systems from the moment of infection. HIV infection satisfies the statutory and regulatory definition of a physical impairment during every stage of the disease.

2. The statute is not operative, and the definition not satisfied, unless the impairment affects a major life activity. Respondent's claim throughout this case has been that the HIV infection placed a substantial limitation on her ability to reproduce and to bear children. App. 14; 912 F. Supp., at 586; 107 F.3d, at 939. Given the pervasive, and invariably fatal, course of the disease, its effect on major life activities of many sorts might have been relevant to our inquiry. Respondent and a number of amici make arguments about HIV's profound impact on almost every phase of the infected person's life. See Brief for Respondent Sidney Abbott 24–27; Brief for American Medical Association as *Amicus Curiae* 20; Brief for Infectious Diseases Society of America et al. as *Amici Curiae* 7–11. In light of these submissions, it may seem legalistic to circumscribe our discussion to the activity

of reproduction. We have little doubt that had different parties brought the suit they would have maintained that an HIV infection imposes substantial limitations on other major life activities.

From the outset, however, the case has been treated as one in which reproduction was the major life activity limited by the impairment. It is our practice to decide cases on the grounds raised and considered in the Court of Appeals and included in the question on which we granted certiorari. See, e.g., *Blessing v. Freestone*, 520 U.S. 329, 340, n. 3 (1997) (citing this Court's Rule 14.1(a)); *Capitol Square Review and Advisory Bd. v. Pinette*, 515 U.S. 753, 760 (1995). We ask, then, whether reproduction is a major life activity.

We have little difficulty concluding that it is. As the Court of Appeals held, "[t]he plain meaning of the word 'major' denotes comparative importance" and "suggest[s] that the touchstone for determining an activity's inclusion under the statutory rubric is its significance." 107 F.3d, at 939, 940. Reproduction falls well within the phrase "major life activity." Reproduction and the sexual dynamics surrounding it are central to the life process itself.

While petitioner concedes the importance of reproduction, he claims that Congress intended the ADA only to cover those aspects of a person's life which have a public, economic, or daily character. Brief for Petitioner 14, 28, 30, 31; see also id., at 36–37 [citing *Krauel v. Iowa Methodist Medical Center*, 95 F.3d 674, 677 (CA8 1996)]. The argument founders on the statutory language. Nothing in the definition suggests that activities without a public, economic, or daily dimension may somehow be regarded as so unimportant or insignificant as to fall outside the meaning of the word "major." The breadth of the term confounds the attempt to limit its construction in this manner.

As we have noted, the ADA must be construed to be consistent with regulations issued to implement the Rehabilitation Act. See 42 U.S.C. § 12201(a). Rather than enunciating a general principle for determining what is and is not a major life activity, the Rehabilitation Act regulations instead provide a representative list, defining term to include "functions such as caring for one's self, performing manual tasks, walking, seeing, hearing, speaking, breathing, learning, and working." 45 CFR § 84.3(j)(2)(ii) (1997); 28 CFR § 41.31(b)(2) (1997). As the use of the term "such as" confirms, the list is illustrative, not exhaustive.

These regulations are contrary to petitioner's attempt to limit the meaning of the term "major" to public activities. The inclusion of activities such as caring for one's self and performing manual tasks belies the suggestion that a task must have a public or economic character in order to be a major life activity for purposes of the ADA. On the contrary, the Rehabilitation Act regulations support the inclusion of reproduction as a major life activity, since reproduction could not be regarded as any less important than working and learning. Petitioner advances no credible basis for confining major life activities to those with a public, economic, or daily aspect. In the absence of any reason to reach a contrary conclusion, we agree with the Court of Appeals' determination that reproduction is a major life activity for the purposes of the ADA.

3. The final element of the disability definition in subsection (A) is whether respondent's physical impairment was a substantial limit on the major life activity she asserts. The Rehabilitation Act regulations provide no additional guidance. 45 CFR pt. 84, App. A, p. 334 (1997).

Our evaluation of the medical evidence leads us to conclude that respondent's infection substantially limited her ability to reproduce in two independent ways. First, a woman infected with HIV who tries to conceive a child imposes on the man a significant risk of becoming infected. The cumulative results of 13 studies collected in a 1994 textbook on AIDS indicates that 20% of male partners of women with HIV became HIV-positive themselves, with a majority of the studies finding a statistically significant risk of infection. Osmond & Padian, "Sexual Transmission of HIV," in *AIDS Knowledge Base* 1.9-8, and tbl. 2; see also Haverkos & Battjes, "Female-to-Male Transmission of HIV," 268 *JAMA* 1855, 1856, tbl. (1992) (cumulative results of 16 studies indicated 25% risk of female-to-male transmission). (Studies report a similar, if not more severe, risk of male-to-female transmission. See, e.g., Osmond & Padian, *AIDS Knowledge Base* 1.9-3, tbl. 1, 1.9-6 to 1.9-7.)

Second, an infected woman risks infecting her child during gestation and childbirth, i.e., perinatal transmission. Petitioner concedes that women infected with HIV face about a 25% risk of transmitting the virus to their children. 107 F.3d, at 942; 912 F. Supp., at 387, n. 6. Published reports available in 1994 confirm the accuracy of this statistic. Report of a Consensus Workshop, *Maternal Factors Involved in Mother-to-Child Transmission of HIV*-1, 5 J. *Acquired Immune Deficiency Syndromes* 1019, 1020 (1992) (collecting 13 studies placing risk between 14% and 40%, with most studies falling within the 25% to 30% range); Connor et al., "Reduction of Maternal-Infant Transmission of Human Immunodeficiency Virus Type 1 with Zidovudine Treatment," 331 *New Eng J Med* 1173, 1176 (1994) (placing risk at 25.5%); see also Strapans & Feinberg, *Medical Management of AIDS* 32 (studies report 13% to 45% risk of infection, with average of approximately 25%).

Petitioner points to evidence in the record suggesting that antiretroviral therapy can lower the risk of perinatal transmission to about 8%. App. 53; see also Connor, supra, at 1176 (8.3%); Sperling et al., "Maternal Viral Load, Zidovudine Treatment, and the Risk of Transmission of Human Immunodeficiency Virus Type 1 from Mother to Infant," 335 *New Eng J Med* 1621,

1622 (1996) (7.6%). The Solicitor General questions the relevance of the 8% figure, pointing to regulatory language requiring the substantiality of a limitation to be assessed without regard to available mitigating measures. Brief for United States as *Amicus Curiae* 18, n. 10 (citing 28 CFR pt. 36, App. B, p. 611 (1997); 29 CFR pt. 1630, App., p. 351 (1997)). We need not resolve this dispute in order to decide this case, however. It cannot be said as a matter of law that an 8% risk of transmitting a dread and fatal disease to one's child does not represent a substantial limitation on reproduction.

The Act addresses substantial limitations on major life activities, not utter inabilities. Conception and childbirth are not impossible for an HIV victim but, without doubt, are dangerous to the public health. This meets the definition of a substantial limitation. The decision to reproduce carries economic and legal consequences as well. There are added costs for antiretroviral therapy, supplemental insurance, and long-term health care for the child who must be examined and, tragic to think, treated for the infection. The laws of some States, moreover, forbid persons infected with HIV from having sex with others, regardless of consent. Iowa Code §§139.1, 139.31 (1997); Md. Health Code Ann. §18-601.1(a) (1994); Mont. Code Ann. §§50-18-101, 50-18-112 (1997); Utah Code Ann. §26-6-3.5(3) (Supp. 1997); id., §26-6-5 (1995); Wash. Rev. Code §9A.36.011(1)(b) (Supp. 1998); see also N. D. Cent. Code §12.1-20-17 (1997).

In the end, the disability definition does not turn on personal choice. When significant limitations result from the impairment, the definition is met even if the difficulties are not insurmountable. For the statistical and other reasons we have cited, of course, the limitations on reproduction may be insurmountable here. Testimony from the respondent that her HIV infection controlled her decision not to have a child is unchallenged. App. 14; 912 F. Supp., at 587; 107 F.3d, at 942. In the context of reviewing summary judgment, we must take it to be true. Fed. Rule Civ. Proc. 56(e). We agree with the District Court and the Court of Appeals that no triable issue of fact impedes a ruling on the question of statutory coverage. Respondent's HIV infection is a physical impairment which substantially limits a major life activity, as the ADA defines it. In view of our holding, we need not address the second question presented, i.e., whether HIV infection is a, *per se*, disability under the ADA.

B

Our holding is confirmed by a consistent course of agency interpretation before and after enactment of the ADA. Every agency to consider the issue under the Rehabilitation Act found statutory coverage for persons with asymptomatic HIV. . . .

[further discussion of various agencies' interpretation of disability and its application to asymptomatic HIV infection is omitted]

The regulatory authorities we cite are consistent with our holding that HIV infection, even in the so-called asymptomatic phase, is an impairment which substantially limits the major life activity of reproduction.

III

The petition for certiorari presented three other questions for review. The questions stated:

"3. When deciding under Title III of the ADA whether a private health care provider must perform invasive procedures on an infectious patient in his office, should courts defer to the health care provider's professional judgment, as long as it is reasonable in light of then-current medical knowledge?

"4. What is the proper standard of judicial review under Title III of the ADA of a private health care provider's judgment that the performance of certain invasive procedures in his office would pose a direct threat to the health or safety of others?

"5. Did petitioner, Randon Bragdon, D.M.D., raise a genuine issue of fact for trial as to whether he was warranted in his judgment that the performance of certain invasive procedures on a patient in his office would have posed a direct threat to the health or safety of others?" Pet. for Cert. i.

Of these, we granted certiorari only on question three. The question is phrased in an awkward way, for it conflates two separate inquiries. In asking whether it is appropriate to defer to petitioner's judgment, it assumes that petitioner's assessment of the objective facts was reasonable. The central premise of the question and the assumption on which it is based merit separate consideration. . . .

[material discussing whether Bragdon's health would have been endangered by treating Abbott is omitted]

We conclude the proper course is to give the Court of Appeals the opportunity to determine whether our analysis of some of the studies cited by the parties would change its conclusion that petitioner presented neither objective evidence nor a triable issue of fact on the question of risk. In remanding the case, we do not foreclose the possibility that the Court of Appeals may reach the same conclusion it did earlier. A remand will permit a full exploration of the issue through the adversary process.

The determination of the Court of Appeals that respondent's HIV infection was a disability under the ADA is affirmed. The judgment is vacated, and the case is remanded for further proceedings consistent with this opinion.

APPENDIX E

EXCERPT FROM *HUMAN CLONING AND HUMAN DIGNITY: AN ETHICAL INQUIRY* EXECUTIVE SUMMARY (JULY 2002)

This is from The President's Council on Bioethics.
[parts omitted]

For the past five years, the prospect of human cloning has been the subject of considerable public attention and sharp moral debate, both in the United States and around the world. Since the announcement in February 1997 of the first successful cloning of a mammal (Dolly the sheep), several other species of mammals have been cloned. Although a cloned human child has yet to be born, and although the animal experiments have had low rates of success, the production of functioning mammalian cloned offspring suggests that the eventual cloning of humans must be considered a serious possibility.

In November 2001, American researchers claimed to have produced the first cloned human embryos, though they reportedly reached only a six-cell stage before they stopped dividing and died. In addition, several fertility specialists, both here and abroad, have announced their intention to clone human beings. The United States Congress has twice taken up the matter, in 1998 and again in 2001–2002, with the House of Representatives in July 2001 passing a strict ban on all human cloning, including the production of cloned human embryos. As of this writing, several cloning-related bills are under consideration in the Senate. Many other nations have banned human cloning, and the United Nations is considering an international convention on the subject. Finally, two major national reports have been issued on human reproductive cloning, one by the National Bioethics Advisory Commission (NBAC) in 1997, the other by the National Academy of Sciences (NAS) in January 2002. Both the NBAC and the NAS reports called for further consideration of the ethical and social questions raised by cloning.

The debate over human cloning became further complicated in 1998 when researchers were able, for the first time, to isolate human embryonic stem cells. Many scientists believe that these versatile cells, capable of becoming any type of cell in the body, hold great promise for understanding and treating many chronic diseases and conditions. Some scientists also believe that stem cells derived from cloned human embryos, produced explicitly for such research, might prove uniquely useful for studying many genetic diseases and devising novel therapies. Public reaction to the prospect of cloning-for-biomedical-research has been mixed: some Americans support it for its medical promise; others oppose it because it requires the exploitation and destruction of nascent human life, which would be created solely for research purposes.

HUMAN CLONING: WHAT IS AT STAKE?

The intense attention given to human cloning in both its potential uses, for reproduction as well as for research, strongly suggests that people do not regard it as just another new technology. Instead, we see it as something quite different, something that touches fundamental aspects of our humanity. The notion of cloning raises issues about identity and individuality, the meaning of having children, the difference between procreation and manufacture, and the relationship between the generations. It also raises new questions about the manipulation of some human beings for the benefit of others, the freedom and value of biomedical inquiry, our obligation to heal the sick (and its limits), and the respect and protection owed to nascent human life.

Finally, the legislative debates over human cloning raise large questions about the relationship between science and society, especially about whether society can or should exercise ethical and prudential control over biomedical technology and the conduct of biomedical research. Rarely has such a seemingly small innovation raised such big questions.

THE INQUIRY: OUR POINT OF DEPARTURE

As Members of the President's Council on Bioethics, we have taken up the larger ethical and social inquiry called for in the NBAC and NAS reports, with the aim of advancing public understanding and informing public policy on the matter. We have attempted to consider human cloning (both for producing children and for biomedical research) within its larger human, technological, and ethical contexts, rather than to view it as an isolated technical development. We focus first on the broad human goods that it may serve as well as threaten, rather than on the immediate impact of the technique itself. By our broad approach, our starting on the plane of human goods, and our open spirit of inquiry, we hope to contribute to a richer and deeper understanding of what human cloning means, how we should think about it, and what we should do about it.

On some matters discussed in this report, Members of the Council are not of one mind. Rather than bury these differences in search of a spurious consensus, we have sought to present all views fully and fairly, while recording our agreements as well as our genuine diversity of perspectives, including our differences on the final recommendations to be made. By this means, we hope to help policymakers and the general public appreciate more thoroughly the difficulty of the issues and the competing goods that are at stake. . . .

THE COUNCIL'S POLICY RECOMMENDATIONS

Having considered the benefits and drawbacks of each of these options, and taken into account our discussions and reflections throughout this report, the Council recommends two possible policy alternatives, each supported by a portion of the Members.

Majority Recommendation: Ten Members of the Council recommend *a ban on cloning-to-produce-children combined with a four-year moratorium on cloning-for-biomedical-research. We also call for a federal review of current and projected practices of human embryo research, pre-implantation genetic diagnosis, genetic modification of human embryos and gametes, and related matters, with a view to recommending and shaping ethically sound policies for the entire field.* Speaking only for ourselves, those of us who support this recommendation do so for some or all of the following reasons:

- By permanently banning cloning-to-produce-children, this policy gives force to the strong ethical verdict against cloning-to-produce-children, unanimous in this Council (and in Congress) and widely supported by the American people. And by enacting a four-year moratorium on the creation of cloned embryos, it establishes an additional safeguard not afforded by policies that would allow the production of cloned embryos to proceed without delay.

- It calls for and provides time for further democratic deliberation about cloning-for-biomedical research, a subject about which the nation is divided and where there remains great uncertainty. A national discourse on this subject has not yet taken place in full, and a moratorium, by making it impossible for either side to cling to the status-quo, would force both to make their full case before the public. By banning all cloning for a time, it allows us to seek moral consensus on whether or not we should cross a major moral boundary (creating nascent cloned human life solely for research) and prevents our crossing it without deliberate decision. It would afford time for scientific evidence, now sorely lacking, to be gathered—from animal models and other avenues of human research—that might give us a better sense of whether cloning-for-biomedical-research would work as promised, and whether other morally nonproblematic approaches might be available. It would promote a fuller and better-informed public debate. And it would show respect for the deep moral concerns of the large number of Americans who have serious ethical objections to this research.
- Some of us hold that cloning-for-biomedical-research can never be ethically pursued, and endorse a moratorium to enable us to continue to make our case in a democratic way. Others of us support the moratorium because it would provide the time and incentive required to develop a system of national regulation that might come into use if, at the end of the four-year period, the moratorium were not reinstated or made permanent. Such a system could not be developed overnight, and therefore even those who support the research but want it regulated should see that at the very least a pause is required. In the absence of a moratorium, few proponents of the research would have much incentive to institute an effective regulatory system. Moreover, the very process of proposing such regulations would clarify the moral and prudential judgments involved in deciding whether and how to proceed with this research.
- A moratorium on cloning-for-biomedical-research would enable us to consider this activity in the larger context of research and technology in the areas of developmental biology, embryo research, and genetics, and to pursue a more comprehensive federal regulatory system for setting and executing policy in the entire area.
- Finally, we believe that a moratorium, rather than a lasting ban, signals a high regard for the value of biomedical research and an enduring concern for patients and families whose suffering such research may help alleviate. It would reaffirm the principle that science can progress while upholding the community's moral norms, and would therefore reaffirm the community's moral support for science and biomedical technology.

The decision before us is of great importance. Creating cloned embryos for *any* purpose requires crossing a major moral boundary, with grave risks and likely harms, and once we cross it there will be no turning back. Our society should take the time to make a judgment that is well-informed and morally sound, respectful of strongly held views, and representative of the priorities and principles of the American people. We believe this ban-plus-moratorium proposal offers the best means of achieving these goals.

This position is supported by Council Members Rebecca S. Dresser, Francis Fukuyama, Robert P. George, Mary Ann Glendon, Alfonso Gómez-Lobo, William B. Hurlbut, Leon R. Kass, Charles Krauthammer, Paul McHugh, and Gilbert C. Meilaender.

Minority Recommendation: Seven Members of the Council recommend *a ban on cloning-to-produce-children, with regulation of the use of cloned embryos for biomedical research.* Speaking only for ourselves, those of us who support this recommendation do so for some or all of the following reasons:

- By permanently banning cloning-to-produce-children, this policy gives force to the strong ethical verdict against cloning-to-produce-children, unanimous in this Council (and in Congress) and widely supported by the American people. We believe that a ban on the transfer of cloned embryos to a woman's uterus would be a sufficient and effective legal safeguard against the practice.
- *It approves cloning-for-biomedical-research and permits it to proceed without substantial delay.* This is the most important advantage of this proposal. The research shows great promise, and its actual value can only be determined by allowing it to go forward now. Regardless of how much time we allow it, no amount of experimentation with animal models can provide the needed understanding of human diseases. The special benefits from working with stem cells from cloned human embryos cannot be obtained using embryos obtained by IVF. We believe this research could provide relief to millions of Americans, and that the government should therefore support it, within sensible limits imposed by regulation.
- It would establish, *as a condition of proceeding*, the necessary regulatory protections to avoid abuses and misuses of cloned embryos. These regulations might touch on the secure handling of embryos, licensing and prior review of research projects, the protection of egg donors, and the provision of equal access to benefits.
- Some of us also believe that mechanisms to regulate cloning-for-biomedical-research should be part of a larger regulatory program governing all research involving human embryos, and that the federal government should initiate a review of present and projected practices of human embryo research, with the aim of establishing reasonable policies on the matter.

Appendix E

Permitting cloning-for-biomedical-research now, while governing it through a prudent and sensible regulatory regime, is the most appropriate way to allow important research to proceed while insuring that abuses are prevented. We believe that the legitimate concerns about human cloning expressed throughout this report are sufficiently addressed by this ban-plus-regulation proposal, and that the nation should affirm and support the responsible effort to find treatments and cures that might help many who are suffering.

This position is supported by Council Members Elizabeth H. Blackburn, Daniel W. Foster, Michael S. Gazzaniga, William F. May, Janet D. Rowley, Michael J. Sandel, and James Q. Wilson.

INDEX

Locators in **boldface** indicate main topics. Locators followed by *g* indicate glossary entries. Locators followed by *b* indicate biographical entries. Locators followed by *c* indicate chronology entries.

Index

Index

Index

Index

Index

Index